MARKOV PROCESSES

Advances in Applied Mathematics

MARKOV PROCESSES

James R. Kirkwood

CRC Press is an imprint of the
Taylor & Francis Group, an **informa** business
A CHAPMAN & HALL BOOK

CRC Press
Taylor & Francis Group
6000 Broken Sound Parkway NW, Suite 300
Boca Raton, FL 33487-2742

Printed on acid-free paper
Version Date: 20150102

International Standard Book Number-13: 978-1-4822-4073-3 (Hardback)

Visit the Taylor & Francis Web site at
http://www.taylorandfrancis.com

and the CRC Press Web site at
http://www.crcpress.com

To Bessie, Katie, and Lizzie—the lights of my life.

Contents

Preface

The purpose of this book is to present a thorough introduction to discrete-state Markov processes. It assumes familiarity with probability and linear algebra at the undergraduate level.

The goal of the writing is to be clear, rigorous, and intuitive. To that end, there is an abundance of computational examples that either immediately precede or follow the theorems they illustrate. Computer algebra systems are expected to be used. Numerical examples are used to develop intuition for why a theorem should be true and should aid in understanding what a theorem is saying in the context of application to a *real-world* problem. It was written so that it would be understandable to someone who has had undergraduate probability and linear algebra courses without needing an instructor to fill in any gaps.

The text was written with several audiences in mind. It provides sufficient background for someone who would like to use Markov processes as a tool for research in a field other than mathematics. It could be used as a text in a second undergraduate level course in probability. A third audience includes someone taking a graduate course in stochastic processes. Such courses often move rapidly, often not giving examples or detailed proofs of the material on Markov processes.

Chapter 1 gives a review of basic probability that will be used in subsequent chapters, emphasizing principles and examples that appear later.

Chapter 2 covers the case of finite-state, discrete-time Markov processes. Preliminary results about stochastic matrices are presented, followed by absorbing processes. A major focus in the analysis of a Markov process is the question of equilibrium state(s). This is introduced and emphasized in this chapter and includes the analyses of reducible and periodic processes in addition to the aperiodic, reducible case. The Perron–Frobenius theorem is used when applicable to demonstrate convergence to equilibrium at an exponential rate.

Chapter 3 deals with the discrete-time, infinite-state case. It begins with renewal processes as they will be fundamental to investigating the equilibrium state of infinite-state processes. Branching processes, in particular the Galton–Watson process, are presented as an application of the infinite-state case.

Chapter 4 provides the background for continuous-time Markov processes with exponential random variables and Poisson processes.

Chapter 5 presents information on continuous-time Markov processes that includes the basic material of Kolmogorov's equations, infinitesimal generators, and explosions. Birth and birth–death processes are developed in detail as is queuing theory. This is where detailed balance equations are introduced.

The final chapter, Chapter 6, covers both discrete and continuous reversible Markov chains.

MATLAB® is a registered trademark of The MathWorks, Inc. For product information, please contact:

The MathWorks, Inc.
3 Apple Hill Drive
Natick, MA 01760-2098 USA
Tel: 508-647-7000
Fax: 508-647-7001
E-mail: info@mathworks.com
Web: www.mathworks.com

1

Review of Probability

The purpose of this text is to give an introduction to Markov chains, which is a branch of mathematics that has applications in a myriad of disciplines including physics, biology, chemistry, economics, and computer science. Markov chains are a subfield of an area of probability called stochastic processes. The text provides an introduction to an area of theoretical mathematics as well as a tool for solving applied problems in diverse fields.

A comment about vocabulary: In the literature, the terms *Markov chains* and *Markov processes* are both used. The difference between them (if, indeed, there is a difference) depends on the author. For some, it refers to whether the state space is discrete or continuous, for some whether time is discrete or continuous, and still others use them interchangeably. We will use them interchangeably.

Short History

A brief historical account of Markov chains could begin with the founding of the Russian Academy of Sciences by Peter the Great in 1724. Initially, many members of the academy came from outside of Russia. Among the prominent mathematicians of the academy in its early years were Leonard Euler and Nicholas and Daniel Bernoulli.

By the nineteenth century, Russian mathematicians occupied a more central role in the academy. Among them was the probabilist Pafnuty Chebyshev, who mentored several students, including Andrei Markov.

Markov was born in 1856 and studied and taught at St. Petersburg University. Markov, Chebyshev, and Jacob Bernoulli (a son of Nicholas Bernoulli) are linked by their research in probability, specifically the law of large numbers.

Markov had a rather cantankerous personality and seemed to have embraced at least some of the Bolshevik philosophies. When Tolstoy was expelled from the Russian church, Markov asked to also be expelled, and when the election of Maxim Gorky to the academy was vetoed by Tsar Nicholas II, Markov refused subsequent honors.

In mathematics, Markov had his strongest disdain for another Russian mathematician, Pavel Nekrasov. Nekrasov tried to mix religious and mathematical arguments, which was anathema to Markov. This led to Markov questioning whether the hypothesis of independence in the law of large numbers could be relaxed in some cases. In 1906, he published his first work on the subject where he examined certain two state systems and expanded his ideas in a 1913 paper that dealt with patterns of letter combinations.

In the 1930s, Andrei Kolmogorov connected these ideas with Brownian motion. Today, Markov chains are used in almost any field that uses mathematical modeling, including some surprising areas such as music.

The setting for our study is we have a finite or countably infinite number of *states* that a *process* can occupy, and the process occupies exactly one of these states at any particular time. The process moves between states, and its movement is governed by a set of conditional probabilities. This distinguishes the situation from many types of probabilistic problems where the next step in a process is independent of the present state. A salient property of this movement is that where the process goes on its next movement depends only on where it is at the present time, and its previous history is of no consequence. When we formally define this last idea, it will be called the Markov property.

Review of Basic Probability Definitions

In this chapter, we present a review of the probability that will be used in our study of Markov chains. It will not cover all areas in an elementary probability course. For example, material from combinatorics is notably absent.

The natural setting of probability is we have an experiment whose outcome depends on chance. We know what the possible outcomes are, and we call this collection of outcomes the sample space.

Examples

1. The experiment is to flip a coin. The possible outcomes are a head or a tail. It is natural to designate these outcomes as H or T. The sample space is usually denoted by S or Ω, so here $\Omega = \{H, T\}$.
2. The experiment is to toss a die. The sample space is $S = \{1,2,3,4,5,6\}$.
3. If we do multiple trials of the same experiment, one often lists the outcomes by the Cartesian product of the outcomes. If the experiment was to flip a coin twice, we could record the outcome as an ordered pair, the first component being the outcome of the first flip and the second component the outcome of the second flip. Thus sample space is $\Omega = \{(H,H),(H,T),(T,H),(T,T)\}$.

This example illustrates the important principle that if an experiment has multiple stages, then the number of possible outcomes of the experiment is the product of the number of outcomes of each stage. So if there are 7 science books, 5 history books, and 3 economics books and we are to choose 1 book from each category, then the number of choices is $7 \times 5 \times 3$.

4. If the experiment is to measure the lifetime of an electrical component, then the set of outcomes is $[0, \infty]$.

Definition

An event is a collection of outcomes to the experiment; that is, an event is a subset of the sample space. We say that an event occurs if the outcome of the experiment is one of the outcomes in the event.

Since events are subsets of the sample space, they can be combined in the usual ways of set theory. The most important of these ways for us are union, intersection, and complementation.

Definition

If A and B are events in the sample space S, then

(i) $A \cup B$ is the event consisting of the outcomes that are in A or B. That is,

$$A \cup B = \{x \mid x \in A \text{ or } x \in B\}.$$

Note that *or* is interpreted in the inclusive sense; that is, $x \in A \cup B$ if $x \in A$ or $x \in B$ or x is in both A and B.

(ii) $A \cap B$ is the event consisting of the outcomes that are in both A and B. That is,

$$A \cap B = \{x \mid x \in A \text{ and } x \in B\}.$$

(iii) A^c (the complement of A) is the set of outcomes that are not in A. So

$$A^c = \{x \in S \mid x \notin A\}.$$

(iv) $A \backslash B$ is the set of outcomes that are in A and not in B. So

$$A \backslash B = A \cap B^c.$$

The ideas of union and intersection are extendible to arbitrary collections of events, so that if $A_1, A_2, A_3, \ldots$ are events, then

$$\cup_i A_i = \{x \mid x \in A_i \quad \text{for some } A_i\}$$

and

$$\cap_i A_i = \{x \mid x \in A_i \quad \text{for every } A_i\}.$$

Definition

A collection of events $A_1, A_2, A_3, \ldots$ is said to be mutually exclusive (or pairwise disjoint) if $A_i \cap A_j = \varnothing$ for $i \neq j$.

Associated with the set of events is a probability measure. Intuitively, the probability measure assigns to each event a number that is the fraction of time the event would be expected to occur if the experiment was performed many times. This number is called the probability of the event. For example, if the experiment is to roll one die, the sample space is {1,2,3,4,5,6}. One event is {1,6}, and the natural way to assign a probability to this event is 2/6 or 1/3.

Definition

A probability measure is a function P defined on the set of subsets of the sample space S that satisfies the following axioms:

1. For any event A, $0 \leq P(A) \leq 1$.
2. $P(S) = 1$.
3. If $A_1, A_2, A_3, \ldots$ is a countable collection of mutually exclusive events, then

$$P(\cup_i A_i) = \sum_i P(A_i).$$

A probability space is the sample space together with the probability measure.

Some Common Probability Distributions

We now describe some of the most important probability distributions.

Bernoulli Distribution

The sample space is {0, 1}. For p, a number with $0 < p < 1$, $P(0) = 1 - p$, and $P(1) = p$. The usual setting for the Bernoulli distribution is we perform an experiment and we classify the outcome as a success or a failure. A success is associated with 1 in the sample space and a failure with 0. It may be the case that there are many outcomes to the experiment, and some of the outcomes are called successes and grouped into the single category of success and the remainder grouped into the single category of failure.

Binomial Distribution

The binomial distribution extends the idea of the Bernoulli distribution in that we perform an experiment and we classify the outcome as a success or a failure, but now we perform the experiment n times and we want to assign a probability to the number of successes in the n trials. The sample space is the possible number of successes in the n trials, so $S = \{0, 1, \ldots, n\}$. For $k \in S$,

$$P(k) = \binom{n}{k} p^k (1-p)^{n-k}$$

where

$$\binom{n}{k} = \frac{n!}{k!(n-k)!}$$

and p is the probability of success on any one trial of the experiment.

To be a probability distribution, we must have

$$\sum_{k=0}^{n} P(k) = \sum_{k=0}^{n} \binom{n}{k} p^k (1-p)^{n-k} = 1.$$

This can be seen to be true from the binomial theorem

$$(a+b)^n = \sum_{k=0}^{n} \binom{n}{k} a^k b^{n-k}$$

by taking $a = p$ and $b = 1 - p$.

Example

We roll a dice and we say that 6 is a success and any other number is a failure. Suppose we roll the dice 10 times and want to compute the probability that exactly 7 of the rolls are a 6.

In the formula, we have

$$p = \frac{1}{6}, \quad 1 - p = \frac{5}{6}, \quad n = 10, \quad k = 7$$

so

$$P(7) = \binom{10}{7}\left(\frac{1}{6}\right)^7\left(\frac{5}{6}\right)^3 = \frac{10!}{7!3!}\left(\frac{1}{6}\right)^7\left(\frac{5}{6}\right)^3 \approx .00025.$$

Geometric Distribution

We conduct an experiment where we classify an outcome as a success or failure. Suppose that p is a number with $0 < p < 1$ and that p is the probability of a success on any one trial of the experiment. We conduct the experiment until the first success. The sample space is the possible number of trials until the first success, so $S = \{1, 2, 3, \ldots\}$. The probability that the first success occurs on the nth trial is

$$(1-p)^{n-1} p.$$

This is because the first $n - 1$ trials had to have been failures and the nth trial a success.

We show $P(S) = 1$.

Now

$$P(S) = \sum_{n=1}^{\infty}(1-p)^{n-1} p = p\sum_{n=1}^{\infty}(1-p)^{n-1} = p\sum_{n=0}^{\infty}(1-p)^n.$$

Using the formula for geometric series

$$\sum_{n=0}^{\infty} r^n = \frac{1}{1-r} \quad \text{if } |r| < 1,$$

we have

$$p\sum_{n=0}^{\infty}(1-p)^n = p\frac{1}{1-(1-p)} = 1.$$

Example

We roll a dice until a 6 appears. So $p = 1/6$ and $1 - p = 5/6$. We compute the probability that the first 6 occurs on the 9th roll. We have

$$P(9) = (1-p)^{9-1} p = \left(\frac{5}{6}\right)^8\left(\frac{1}{6}\right) \approx .039.$$

Negative Binomial Distribution

We have a sequence of Bernoulli trials, each having a probability of success p, $0 < p < 1$. The negative binomial distribution is the probability distribution for the number of trials required to achieve r successes. The probability that if n is the number of trials required to achieve r successes, then

$$\binom{n-1}{r-1} p^r (1-p)^{n-r}.$$

The reasoning for this formula is that for the rth success to occur on the nth trial, there must be $r - 1$ successes in the first $n - 1$ trials and then a success on the nth trial. From the binomial distribution, we know the probability of $r - 1$ successes in the first $n - 1$ trials is

$$\binom{n-1}{r-1} p^{r-1} (1-p)^{(n-1)-(r-1)} = \binom{n-1}{r-1} p^{r-1} (1-p)^{n-r}.$$

We then need a success on the nth trial, which occurs with probability p. Thus, the probability that exactly n trials are required to attain r successes is

$$p\binom{n-1}{r-1} p^{r-1} (1-p)^{n-r} = \binom{n-1}{r-1} p^r (1-p)^{n-r}.$$

We need to show

$$\sum_{n=1}^{\infty} \binom{n-1}{r-1} p^r (1-p)^{n-r} = 1,$$

which is Exercise 1.10.

Example

We roll a fair die until we get three 6s.

What is the probability that exactly five rolls will be required?

We now have $p = 1/6$, $r = 3$, and $n = 5$, so the probability exactly five rolls will be required is

$$\binom{5-1}{3-1}\left(\frac{1}{6}\right)^3\left(1-\frac{1}{6}\right)^{5-3} = 6\left(\frac{1}{6}\right)^3\left(\frac{5}{6}\right)^2 \approx .193.$$

Poisson Distribution

In the Poisson distribution, we monitor the occurrence of an event. One knows the average rate at which the event occurs, and subsequent events occur independently of previous occurrences of the event. (We will later refer to this last characteristic as being memoryless.) For a given fixed period of time, we assign a probability to the number of times that the event occurs during that period. The sample space is the possible number of occurrences, so $S = \{0, 1, 2, 3, \ldots\}$. If $\lambda > 0$ is the rate of an occurrence, and k is the number of occurrences in the time period, then we set

$$P(k) = e^{-\lambda}\frac{\lambda^k}{k!}.$$

Note that

$$P(S) = \sum_{k=0}^{\infty} P(k) = \sum_{k=0}^{\infty} e^{-\lambda}\frac{\lambda^k}{k!} = e^{-\lambda}\sum_{k=0}^{\infty}\frac{\lambda^k}{k!} = e^{-\lambda}e^{\lambda} = 1.$$

Example

A bank teller on the average has 8 customers per hour. What is the probability that she will have 5 or fewer customers in an hour?

Here $\lambda = 8$, so the probability of 5 or fewer customers is

$$\sum_{k=0}^{5} e^{-8}\frac{8^k}{k!} \approx .191.$$

Properties of a Probability Distribution

Several other properties of a probability measure are derivable from the axioms. The first theorem gives some of these.

Theorem 1.1

If A and B are events and S is the sample space, then

(i) $P(\varnothing) = 0$
(ii) $P(A^c) = 1 - P(A)$
(iii) if $A \subset B$, then $P(A) \leq P(B)$
(iv) $P(A \cup B) = P(A) + P(B) - P(A \cap B)$

Proof

(i) We have

$$P(A)=P(A\cup\varnothing)=P(A)+P(\varnothing)$$

so $P(\varnothing) = 0$.

(ii) We have

$$1=P(S)=P(A\cup A^c)=P(A)+P(A^c)$$

so $P(A^c) = 1 - P(A)$.

(iii) If $A\subset B$, we have

$$B=A\cup(B\backslash A) \quad \text{and} \quad A\cap(B\backslash A)=\varnothing$$

so

$$P(B)=P\left[A\cup(B\backslash A)\right]=P(A)+P\left[A\cup(B\backslash A)\right],$$

and since $P\left[A\cup(B\backslash A)\right]\geq 0$, we have $P(A) \leq P(B)$.

(iv) We have

$$A\cup B=(A\backslash B)\cup(B\backslash A)\cup(A\cap B),$$

and the sets on the right are disjoint. So

$$P(A\cup B)=P(A\backslash B)+P(B\backslash A)+P(A\cap B).$$

Then

$$P(A\cup B)+P(A\cap B)=\left[P(A\backslash B)+P(A\cap B)\right] +\left[P(B\backslash A)+P(A\cap B)\right]=P(A)+P(B)$$

so

$$P(A\cup B)=P(A)+P(B)-P(A\cap B).$$

Conditional Probability

The idea with conditional probability is that in determining the likelihood of an event, we have some additional information that something has already occurred. That additional information may or may not change the likelihood the event will occur.

Example

Someone draws a card from a deck, and we want to assign a probability to some characteristic of the drawn card. We have a friend who caught a glimpse of the drawn card and informs us the card was red.

1. If we are to assign a probability to the event *the card drawn was an ace*, then the additional information does not change the probability the event will occur. This is because in a 52-card deck there are 4 aces, so the probability that an ace is drawn is

   $$\frac{4}{52} = \frac{1}{13}.$$

   Once we know that the drawn card was red, we have eliminated 26 of the possible outcomes; that is, the black cards. In a sense, we have decreased the sample space. Now we look just at the red cards. There are 26 of these, 2 of which are aces. Thus, the probability of having drawn an ace is

   $$\frac{2}{26} = \frac{1}{13}$$

   so the additional knowledge that a red card was drawn does not affect the probability the card drawn was an ace.
2. If we are to assign a probability to the event *the card drawn was a heart*, then the probability of whether the event occurred is altered. In a 52-card deck, there are 13 hearts, so the probability that a heart will be drawn is

   $$\frac{13}{52} = \frac{1}{4}.$$

   If we know the card drawn was red, then we have again eliminated 26 of the possible outcomes and we consider only the 26 red cards. Of these, 13 are hearts, so the probability that a heart will be drawn is

   $$\frac{13}{26} = \frac{1}{2}$$

   and the additional knowledge has altered the probability that the event the card drawn was an heart.

   If A and B are events, we denote the probability that A occurs, given that B has occurred by $P(A|B)$. This is read *the probability of A given B.*

   The reasoning for the formula to compute $P(A|B)$ is as follows: If B has occurred, then for A to also occur, we must have that both A and B occur; that is, the event $A \cap B$ occurs. Once we know B has occurred, we have reduced the sample space to B, so what is to be computed is the probability of $A \cap B$ relative to B. Thus,

$$P(A|B) = \frac{P(A \cap B)}{P(B)}, \quad (1.1)$$

which is meaningful only if $P(B) \neq 0$.

Example

We flip two fair coins. What is the probability both are heads, given that one is a head?

Solution

The sample space for flipping two coins is $\{(H,H),(H,T),(T,H),(T,T)\}$.

Let A be the event both are heads. Then $A = \{(H,H)\}$ and $P(A) = 1/4$.

Let B be the event at least one is a head. Then $B = \{(H,H)(H,T),(T,H)\}$ and $P(B) = 3/4$.

In this case, $A \cap B = A$, so $P(A \cap B) = P(A) = 1/4$.

Then the probability both are heads, given that one is a head, is

$$P(A|B) = \frac{P(A \cap B)}{P(B)} = \frac{1/4}{3/4} = \frac{1}{3}.$$

Independent Events

As the words suggest, A and B are independent events if the occurrence of one has no effect on the probability that the other will occur. So $P(A|B) = P(A)$ and $P(B|A) = P(B)$. In terms of Equation 1.1, this says

$$P(A) = P(A|B) = \frac{P(A \cap B)}{P(B)}$$

or

$$P(A \cap B) = P(A)P(B).$$

This is the definition of two events being independent.

Definition

The events A and B are independent if $P(A \cap B) = P(A)P(B)$. A collection of events $E_1, E_2, E_3, \ldots, E_n$ is said to independent if

$$P(E_{i1} \cap E_{i2} \cap \cdots \cap E_{ik}) = P(E_{i1})P(E_{i2})\cdots P(E_{ik})$$

for any collection of distinct sets in $\{E_1, E_2, E_3, \ldots, E_n\}$.

A collection of events $E_1, E_2, E_3, \ldots, E_n$ is said to pairwise independent if

$$P(E_i \cap E_j) = P(E_i)P(E_j)$$

for any two distinct sets in $\{E_1, E_2, E_3, \ldots, E_n\}$.

Example

This example shows that pairwise independence of a collection of sets does not ensure independence of the collection.

Let $S = \{1,2,3,4\}$ and $A = \{1,2\}, B = \{1,3\}, C = \{1,4\}$.

If the experiment is to draw a number from S and all numbers have equal probability, then

$$P(A) = P(B) = P(C) = \frac{1}{2}$$

and

$$A \cap B = A \cap C = B \cap C = \{1\}, \quad \text{so } P(A \cap B) = P(A \cap C) = P(B \cap C) = \frac{1}{4}.$$

Then

$$P(A \cap B) = \frac{1}{4} = P(A)P(B), \quad P(A \cap C) = \frac{1}{4} = P(A)P(C),$$

$$P(B \cap C) = \frac{1}{4} = P(B)P(C),$$

and A, B, and C are pairwise independent.

But

$$A \cap B \cap C = \{1\} \quad \text{so } P(A \cap B \cap C) = \frac{1}{4} \text{ and } P(A)P(B)P(C) = \frac{1}{2} \cdot \frac{1}{2} \cdot \frac{1}{2} = \frac{1}{8}$$

so A, B, and C are not independent.

Random Variables

Definition

A random variable is a function from the sample space to the real numbers.

In other words, a random variable associates a real number with each outcome of the experiment. Random variables are typically denoted X, Y, or Z in contrast to other fields of mathematics where functions are usually denoted by names such as f, g, or h.

Definition

A random variable is said to be discrete if it takes on at most countably many values. (This includes finitely many values.)

Discrete random variables are the random variables that we consider in the remainder of this chapter. The other type of random variable that we will consider (continuous random variables) will not be pertinent until Chapter 4, and we will defer our discussion of continuous random variables until then.

Example

The experiment is to roll two dice. There are 36 possible outcomes,

$$\begin{gathered}(1,1),(1,2),\ldots,(1,6)\\(2,1),(2,2),\ldots,(2,6)\\\vdots\\(6,1),(6,2),\ldots,(6,6).\end{gathered}$$

In a board game such as Monopoly®, a player rolls a pair of dice and then moves his token the number of spaces that is the sum of the dice. It is natural in this case to create a random variable that assigns to a roll of the dice the sum of the two dice. So if X is this random variable, then $X((6,2)) = 8$.

It is often the case, as it is in this example, that an important piece of information is how likely is it that the random variable has a certain value. If we need to move 8 spaces to land on a particular square, we might want to know the probability of rolling an 8.

It is necessary to become comfortable with statements such as $P(X = 8)$. Once one becomes accustomed to such statements, they are very intuitive but the notation has some nuances that are different from other areas of mathematics, and we digress for a bit in order to clarify some notational points. If one feels comfortable with the statement $P(X = 8)$, then the next explanation can be skipped.

Suppose we have a function f from the set X to the set Y

$$f: X \to Y.$$

Associated with each such function f, there is a function f^{-1} that takes *subsets of* Y to *subsets of* X. This function is defined as follows: If $B \subset Y$, then

$$f^{-1}(B) = \{x \in X \mid f(x) \in B\}.$$

While the symbol f^{-1} is identical to that used in precalculus, this has nothing to do with the function f^{-1} of precalculus unless f is a one-to-one function.

Another way to describe $f^{-1}(B)$ is

$$x \in f^{-1}(B) \quad \text{if and only if} \quad f(x) \in B.$$

Example

If $f(x) = x^2$ then

$$f^{-1}(\{4,9\}) = \{-2,2,-3,3\}, \quad f^{-1}([1,16]) = [-4,-1] \cup [1,4]$$

$$f^{-1}([-12,-6]) = \varnothing.$$

In probability, we often have a situation where X is a random variable and we are asked to consider an expression such as $P(X = 8)$. Here, X is a function that maps the sample space into the real numbers

$$X: S \to \mathbb{R},$$

and with the preceding notation, X^{-1} would map subsets of $\mathbb{R}$ to subsets of S. In the dice rolling example, $X^{-1}(\{8\}) = \{(2,6),(3,5),(4,4),(5,3),(6,2)\}$ and the probability of the event $\{(2,6),(3,5),(4,4),(5,3),(6,2)\}$ is 5/36. Thus, in most mathematical symbolism, we would say $P(X^{-1}(\{8\})) = 5/36$. In probability, one phrases this in the more efficient and intuitive way $P(X = 8) = 5/36$.

Associated with each random variable X are two functions, the probability density function (p.d.f.) that is usually denoted $p(x)$ in the discrete case and $f(x)$ in the continuous case and the cumulative distribution function (c.d.f.) that is denoted $F(x)$ in both cases. In the discrete case, these are defined by

$$p(x) = P(X = x) \quad \text{and} \quad F(x) = P(X \le x).$$

Expected Value of a Random Variable

Several parameters are associated with a random variable, the most prominent being the expected value and the variance. In our study, the expected value will be used most often. To introduce the expected value, we consider the following example.

Example

Suppose scores on five tests are

$$80, 90, 80, 70, 90.$$

To compute the average, we would sum the scores and divide by 5. That is

$$\frac{80+90+80+70+90}{5}.$$

This can be expressed in several equivalent ways, such as

$$\frac{80+90+80+70+90}{5}=\frac{70+80+80+90+90}{5}$$
$$=\frac{1\times 70+2\times 80+2\times 90}{5}=70\times\frac{1}{5}+80\times\frac{2}{5}+90\times\frac{2}{5}.$$

The last way to compute the average is significant for our purposes. We took each score, multiplied it by the fraction of time it occurred, and summed the products. If x is a score and $p(x)$ is the fraction of time the score occurred, we can compute the average, $\bar{x}$, with the formula

$$\bar{x}=\sum_{x} x\,p(x)$$

where the sum is over the scores that occurred.

This has an advantage in that we do not need to know the number of times a score occurred, but only the fraction of time—or probability—that a score occurred.

In this example, we computed a weighted average. This is the way the expected valued of a discrete random variable is computed.

Definition

Let S be the sample space and X be a discrete random variable; that is, the range of X is a countable set of real numbers, $n_1, n_2, \ldots$.

Let $A_i=\{x\in S|X(s)=n_i\}$. (Another way to describe A_i is $A_i = X^{-1}(n_i)$.)

Then the sets A_i form a partition of the sample space S; that is,

$$S=\cup A_i \quad \text{and} \quad A_i\cap A_j=\varnothing \quad \text{if } i\neq j.$$

The expected value of the discrete random variable X, denoted $E[X]$, is

$$E[X]=\sum_{i} n_i P(A_i)=\sum_{i} n_i P(X=n_i)$$

provided that

$$\sum_i |n_i| P(X = n_i)$$

converges. If the last series does not converge, then X has no finite expected value.

This formula is usually expressed as

$$E[X] = \sum_x xp(x)$$

where the sum is over the values of x that are assumed by X.

Example

We roll a pair of dice, and let X be the sum of the dice. Then X takes on the values {2,3,4,…,12}. We have

$$P(X=2) = P\{(1,1)\} = \frac{1}{36}$$

$$P(X=3) = P\{(1,2),(2,1)\} = \frac{2}{36}$$

$$P(X=4) = P\{(1,3),(2,2),(3,1)\} = \frac{3}{36}$$

$$P(X=5) = P\{(1,4),(2,3),(3,2),(4,1)\} = \frac{4}{36}$$

$$P(X=6) = P\{(1,5),(2,4),(3,3),(4,2),(5,1)\} = \frac{5}{36}$$

$$P(X=7) = P\{(1,6),(2,5),(3,4),(4,3),(5,2),(6,1)\} = \frac{6}{36}$$

$$P(X=8) = P\{(2,6),(3,5),(4,4),(5,3),(6,2)\} = \frac{5}{36}$$

$$P(X=9) = P\{(3,6),(4,5),(5,4),(6,3)\} = \frac{4}{36}$$

$$P(X=10) = P\{(4,6),(5,5),(6,4)\} = \frac{3}{36}$$

$$P(X=11) = P\{(5,6),(6,5)\} = \frac{2}{36}$$

$$P(X=12) = P\{(6,6)\} = \frac{1}{36}$$

and

$$E[X]=2\left(\frac{1}{36}\right)+3\left(\frac{2}{36}\right)+4\left(\frac{3}{36}\right)+5\left(\frac{4}{36}\right)+6\left(\frac{5}{36}\right)+7\left(\frac{6}{36}\right)$$
$$+8\left(\frac{5}{36}\right)+9\left(\frac{4}{36}\right)+10\left(\frac{3}{36}\right)+11\left(\frac{2}{36}\right)+12\left(\frac{1}{36}\right)=\frac{126}{36}=3\frac{1}{2}.$$

The interpretation of $E[X]$ is that if we kept track of many rolls of the dice, we expect the average of the rolls would be close to 3.5.

Properties of the Expected Value

The most basic properties of expected value are given by the next theorem.

Theorem 1.2

(a) If X is a random variable and c is a constant, then $E[X + c] = E[X] + c$.
(b) If X and Y are random variables defined on the same probability space, then $E[X + Y] = E[X] + E[Y]$.
(c) If X is a random variable and c is a constant, then $E[cX] = cE[X]$.

The proof is left as an exercise.

Expected Value of a Random Variable with Common Distributions

Bernoulli Distribution

For a random variable X with a Bernoulli distribution, we have

$$P(X=0)=1-p,\quad P(X=1)=p$$

so

$$E[X]=0(1-p)+1\cdot p=p.$$

Binomial Distribution

Suppose X is a random variable that has a binomial distribution of n-independent trials, each trial a Bernoulli trial with probability of success p. If Y_i is defined by

$$Y_i = \begin{cases} 0 & \text{if the } i\text{th trial is a failure} \\ 1 & \text{if the } i\text{th trial is a success}' \end{cases}$$

then Y_i is a Bernoulli random variable with probability of success p, and

$$E[X] = E[Y_1 + Y_2 + \cdots + Y_n] = E[Y_1] + E[Y_2] + \cdots + E[Y_n] = np.$$

Geometric Distribution

We perform an experiment that has two outcomes that are called success and failure. The trials are independent, and we want to know how many trials are likely until the first success. We let

$$p = \text{probability of success}, \quad \text{so } (1-p) \text{is the probability of failure.}$$

If X is the random variable that gives the number of the rolls until we get the first success, then the first success occurs on the nth trial if the first $(n - 1)$ trials are failures and the nth trial is a success. So $P(X = n) = (1 - p)^{n-1}p$ and

$$E[X] = \sum_{n=1}^{\infty} n(1-p)^{n-1}p. \tag{1.2}$$

To compute

$$\sum_{n=1}^{\infty} n(1-p)^{n-1}p = p\sum_{n=1}^{\infty} n(1-p)^{n-1},$$

consider that for $0 < p < 1$

$$\sum_{n=0}^{\infty} (1-p)^n = \frac{1}{1-(1-p)} = \frac{1}{p}. \tag{1.3}$$

Since the series in Equation 1.3 converges uniformly on any compact interval $[\epsilon, 1 - \epsilon]$, it is legitimate to say

$$\frac{d}{dp}\sum_{n=0}^{\infty}(1-p)^n = \frac{d}{dp}\left(\frac{1}{p}\right) \quad \text{for } 0 < p < 1$$

so

$$\sum_{n=0}^{\infty} n(1-p)^{n-1}(-1) = -\frac{1}{p^2}$$

and

$$p\sum_{n=0}^{\infty} n(1-p)^{n-1} = p\left(\frac{1}{p^2}\right) = \frac{1}{p}.$$

A random variable as previously defined is called a geometric random variable. It is worthwhile to remember that for such a random variable

$$E[X] = \frac{1}{p}.$$

Example

We roll a dice until a 6 appears. Let X be the random variable that gives the number of the rolls until we get the first 6. Then

$$p = \frac{1}{6}$$

so

$$E[X] = \frac{1}{1/6} = 6.$$

Negative Binomial Distribution

Recall that in the negative binomial distribution, we have a sequence of Bernoulli trials, each having a probability of success p, $0 < p < 1$. The negative binomial distribution is the probability distribution for the number of trials required to achieve r successes. If X_r is the random variable that gives the number of trials required to achieve r successes, then

$$P(X_r = n) = \binom{n-1}{r-1} p^r (1-p)^{n-r}.$$

To find the expected value of X_r, let Y_k be the random variable that gives the number of trials between the $(k-1)$st and kth success. Then $Y_1, Y_2, \ldots, Y_r$ are independent identically distributed geometric random variables with

$$E[Y_k] = \frac{1}{p}$$

and

$$X_r = Y_1 + Y_2 + \cdots + Y_r.$$

Thus,

$$E[X_r] = E[Y_1 + Y_2 + \cdots + Y_r] = E[Y_1] + \cdots + E[Y_r] = r\frac{1}{p} = \frac{r}{p}.$$

Example

We roll a fair die until we get three 6s. Find the expected number of rolls needed.

We have $p = 1/6$, so $1/p = 6$ and $r = 3$, so the expected number is

$$6 \times 3 = 18.$$

Poisson Distribution

For the Poisson distribution, we have

$$P(X = n) = e^{-\lambda} \frac{\lambda^n}{n!}, \quad \lambda > 0, n = 0, 1, 2, \ldots$$

so

$$E[X] = \sum_{n=0}^{\infty} n e^{-\lambda} \frac{\lambda^n}{n!} = \sum_{n=1}^{\infty} n e^{-\lambda} \frac{\lambda^n}{n!} = e^{-\lambda} \sum_{n=1}^{\infty} \frac{\lambda^n}{(n-1)!} = \lambda e^{-\lambda} \sum_{n=1}^{\infty} \frac{\lambda^{n-1}}{(n-1)!}$$

$$= \lambda e^{-\lambda} \sum_{k=0}^{\infty} \frac{\lambda^k}{k!} = \lambda.$$

Functions of a Random Variable

Suppose that $f(x)$ is a function from the real numbers to the real numbers, say $f(x) = x^2$. If X is a random variable, then we can form the random variable $f(X)$. In this case,

$$f(X) = X^2.$$

Note that $f(X)$ is the composition of the functions $X: S \to \mathbb{R}$ followed by

$$f: \mathbb{R} \to \mathbb{R}.$$

We compute the expected value of $f(X)$; that is, $E[f(X)]$.

Using the previous notation, for n_i in the range of X, $f(n_i)$ is in the range of $f(X)$. The probability of n_i is $P(A_i)$ so the contribution to $E[f(X)]$ from the points of the sample space that are in A_i is $f(n_i)P(A_i)$.

Summing over i, we get

$$E[f(X)] = \sum_i f(n_i)P(A_i) = \sum_i f(n_i)P(X = n_i),$$

which is usually expressed as

$$E[f(X)] = \sum_x f(x)p(x).$$

(Note that we could have $f(n_i) = f(n_j)$ for $n_i \neq n_j$.)

This result is formalized by the following theorem.

Theorem 1.3

If X is a discrete random variable with p.d.f. $p(x)$ and $f: \mathbb{R} \to \mathbb{R}$, then

$$E[f(X)] = \sum_x f(x)p(x).$$

Example

Suppose $p(1) = .2$, $p(3) = .1$ and $p(6) = .7$. We compute $E[X]$ and

$$E[3X^2 + 1].$$

We have

$$E[X] = 1 \times .2 + 3 \times .1 + 6 \times .7 = 4.7$$

and

$$E[3X^2 + 1] = [3(1^2) + 1] \times .2 + [3(3^2) + 1] \times .1 + [3(6^2) + 1] \times .7 = 79.9.$$

Corollary

If X is a random variable and a and b are real numbers, then

$$E[aX + b] = aE[X] + b.$$

Proof

This follows from Theorem 1.2.

The next result is an alternate way to find the expected value of certain random variables.

Theorem 1.4

If Y is a discrete random variable whose range is in $\{1, 2, \ldots\}$, then

$$E[Y] = \sum_{n=1}^{\infty} P(Y \geq n).$$

Proof

We have

$$E[Y] = \sum_{n=1}^{\infty} nP(Y = n) = 1 \cdot P(Y = 1) + 2 \cdot P(Y = 2) + 3 \cdot P(Y = 3) + \cdots,$$

which can be expressed as

$$\begin{array}{c} P(Y=1)+ \\ P(Y=2) + P(Y=2)+ \\ P(Y=3) + P(Y=3) + P(Y=3)+ \\ \vdots \end{array}$$

Sum the columns to get

$$E[Y] = P(Y \geq 1) + P(Y \geq 2) + P(Y \geq 3) + \cdots = \sum_{n=1}^{\infty} P(Y \geq n).$$

In fact, the proof consisted of interchanging the order of two sums.

The next result will be used repeatedly in our study of Markov chains.

Theorem 1.5 (Law of Total Probability)

If A, B_1, B_2, B_3, … are events with $\{B_1, B_2, B_3, \ldots\}$ mutually exclusive and $\cup_i B_i = S$, then

$$P(A) = \sum_i P(A|B_i)P(B_i).$$

Proof
We have

$$A = A \cap S = A \cap (\cup_i B_i) = (A \cap B_1) \cup (A \cap B_2) \cup (A \cap B_3) \cup \cdots.$$

Since $B_i \cap B_j = \varnothing$ if $i \neq j$, then $(A \cap B_i) \cap (A \cap B_j) = \varnothing$ if $i \neq j$. Thus,

$$P(A) = \sum_i P(A \cap B_i).$$

But

$$P(A|B_i) = \frac{P(A \cap B_i)}{P(B_i)},$$

so

$$P(A|B_i)P(B_i) = P(A \cap B_i)$$

and thus

$$P(A) = \sum_i P(A|B_i)P(B_i).$$

Example

We have two urns. The first urn contains 8 red balls and 4 green balls. The second contains 1 red ball and 4 green balls. We select an urn at random, and then choose a ball from the selected urn. What is the probability the ball chosen is red?

Solution

Let B_i be the event that the ith urn is selected. Then $P(B_i) = 1/2$.
Let A be the event that a red ball is chosen. Then $P(A|B_1) = 8/12$ and $P(A|B_2) = 1/5$.

So

$$P(A) = P(A|B_1)P(B_1) + P(A|B_2)P(B_2) = \frac{8}{12}\cdot\frac{1}{2} + \frac{1}{5}\cdot\frac{1}{2} = \frac{13}{30}.$$

Note that there are a total of 9 red balls and 8 green balls. If the balls were mixed together and a ball were drawn, the probability the ball would be red is 9/17.

Joint Distributions

In many situations, one would like to simultaneously consider two or more properties of the outcome of an experiment. For example, if we wanted to understand the physical profile of a group of people, we might select a member of the population and measure that person's height and weight. In this case, the sample space would be the population from which the person is chosen, and the outcome of the experiment would be the person chosen. We could let X be the height of the person chosen and Y their weight.

Definition

If X and Y are discrete random variables, their joint p.d.f. is

$$p(x,y) = P(X = x, Y = y).$$

Notation: $P(X = x, Y = y)$ means the same as $P\big((X = x) \cap (Y = y)\big)$.

A common notation for the joint p.d.f. in the continuous case is $f(x,y)$.

For the case of two discrete random variables, it is convenient to express the joint p.d.f. as a matrix. Since $p(x,y)$ is a probability, we have

$$p(x,y) \geq 0 \quad \text{and} \quad \sum_x \left(\sum_y p(x,y) \right) = 1.$$

From the joint p.d.f. of X and Y, we can recover the p.d.f. of each random variable. If $p(x,y)$ is the joint p.d.f. of X and Y, then

$$p_X(x) = P(X = x) = \sum_y p(x,y) \quad \text{and} \quad p_Y(y) = P(Y = y) = \sum_x p(x,y)$$

are the p.d.f.'s of X and Y, respectively. These are called the marginal p.d.f.'s of the random variables.

Example

Consider the joint p.d.f. expressed in Table 1.1.
For the data in Table 1.1,

$$p_X(1) = .33, \quad p_X(2) = .31, \quad p_X(3) = .36$$

$$p_Y(1) = .24, \quad p_Y(2) = .25, \quad p_Y(3) = .31, \quad p_Y(4) = .20.$$

Definition

The random variables X and Y are independent if for any sets of real numbers A and B

$$P(X \epsilon A, Y \epsilon B) = P(X \epsilon A)P(Y \epsilon B).$$

This is true if and only if $p(x,y) = p_X(x)p_Y(y)$ where $p(x,y)$ is the joint p.d.f. of X and Y and $p_X(x)$ and $p_Y(y)$ are the p.d.f.'s of X and Y, respectively.

TABLE 1.1

Joint Probability Density Function

	X = 1	**X = 2**	**X = 3**	
$Y = 1$	$P(1,1) = .07$	$P(2,1) = .06$	$P(3,1) = .11$	$\sum P(x,1) = .24$
$Y = 2$	$P(1,2) = .11$	$P(2,2) = .08$	$P(3,2) = .06$	$\sum P(x,2) = .25$
$Y = 3$	$P(1,3) = .09$	$P(2,3) = .13$	$P(3,3) = .09$	$\sum P(x,3) = .31$
$Y = 4$	$P(1,4) = .06$	$P(2,4) = .04$	$P(3,4) = .10$	$\sum P(x,4) = .20$
	$\sum P(1,y) = .33$	$\sum P(2,y) = .31$	$\sum P(3,y) = .36$	

Theorem 1.6

If X and Y are random variables, then

$$E[X+Y]=E[X]+E[Y].$$

Proof

This result is similar to one noted previously. The difference is that here there are different p.d.f.'s, one for X and one for Y.

Suppose that $p(x,y)$ is the joint p.d.f. for X and Y. We have

$$E[X+Y]=\sum_{x,y}(x+y)p(x,y)=\sum_{x,y}xp(x,y)+\sum_{x,y}yp(x,y)$$

$$=\sum_{x}\left(\sum_{y}xp(x,y)\right)+\sum_{y}\left(\sum_{x}yp(x,y)\right)$$

$$=\sum_{x}x\left(\sum_{y}p(x,y)\right)+\sum_{y}y\left(\sum_{x}p(x,y)\right)$$

$$=\sum_{x}xp_X(x)+\sum_{y}yp_Y(y)=E[X]+E[Y].$$

Corollary

If X and Y are random variables and a and b are real numbers, then

$$E[aX+bY]=aE[X]+bE[Y].$$

Theorem 1.7

If X and Y are independent random variables, then

$$E[XY]=E[X]E[Y].$$

Proof

We have

$$E[XY]=\sum_{x,y}xyp(x,y)=\sum_{x,y}xyp(x)p(y)=\sum_{x}xp(x)\sum_{y}yp(y)=E[X]E[Y].$$

Corollary

If X and Y are independent random variables and f and g are functions from $\mathbb{R}$ to $\mathbb{R}$, then

$$E\left[f(X)g(Y)\right] = E\left[f(X)\right]E\left[g(Y)\right].$$

Convolution

For $f(x)$ and $g(x)$ real-valued integrable functions, the convolution of f and g, denoted $f * g$, is defined by

$$(f * g)(y) = \int_{-\infty}^{\infty} f(y - x)g(x)\,dx.$$

It can be shown that $f * g = g * f$.

The analog for discrete integer valued functions is

$$(f * g)(n) = \sum_{k=-\infty}^{\infty} f(n-k)g(k).$$

An important use of convolution is the probability distribution of the sum of independent random variables.

Theorem 1.8

Suppose that X and Y are independent discrete integer-valued random variables with p.d.f.'s f and g, respectively. The p.d.f. of $X + Y$ is $h(n)$ where

$$h(n) = (f * g)(n) = \sum_{k \in \mathbb{Z}} f(n-k)g(k) = \sum_{k \in \mathbb{Z}} P(X = n-k)P(Y = k).$$

Proof

We have

$$P(X+Y=n) = \sum_{k} P(X = n-k, Y = k) = \sum_{k} P(X = n-k)P(Y = k) = \sum_{k} f(n-k)g(k)$$

where the second equality is valid because of independence.

Example

Previously, we considered finding the probability that the sum of the rolls of a pair of dice was a particular value. Here we consider the same problem from the point of view of a convolution. Suppose we roll two dice, say a red die and a green die, and we want to find the probability that the sum of the dice is 8.

Let X be the outcome of the red die and Y be the outcome of the green die. Let f and g be the p.d.f.'s of X and Y, respectively. We could compute $P(X + Y = 8)$ using

$$P(X+Y=8)=(f*g)(n)=\sum_{n=2}^{6}\left(P(X=n)\right)\left(P(Y=8-n)\right)$$

$$=\sum_{n=2}^{6}\left(\frac{1}{6}\right)\left(\left(\frac{1}{6}\right)\right)=5\left(\frac{1}{36}\right)$$

where the sum is from 2 to 6 because each die must have a value of at least 2 in order for the sum to be 8. Of course, this gives the same answer as the earlier computation.

Corollary

If $X_1, X_2, \ldots, X_n$ are independent, identically distributed random variables with p.d.f. f, then the p.d.f. of $X_1 + X_2 + \cdots + X_n$ is $f*f*\cdots*f$ where

$$f*f*f=f*(f*f).$$

Theorem 1.9

If X and Y are independent and identically distributed discrete random variables that take values in the nonnegative integers with common density distribution f and cumulative distribution F, then $X + Y$ has the cumulative distribution

$$(F*f)(n)=\sum_{k=0}^{n}F(n-k)f(k).$$

The proof is left as Exercise 1.37.

Generating Functions

Generating functions provide an efficient computational tool for certain characteristics of random variables, including the expectation and variance. The generating functions we will study are the probability generating function and the moment generating function.

Probability Generating Functions

The probability generating function is used in the case where the random variable takes values in the nonnegative integers. If X is such a random variable, then the probability generating function for X, denoted $G(z)$, is defined by

$$G(z) = E\left(z^X\right) = \sum_{n=0}^{\infty} p(n) z^n = p(0) + p(1)z + p(2)z^2 + \cdots.$$

where $p(n) = P(X = n)$.

If the random variable X is not evident, then we write $G_X(z)$.

We will use

$$\frac{d}{dz} G(z)\Big|_{z=1} = \sum_{n=0}^{\infty} np(n) z^{n-1}\Big|_{z=1} = \sum_{n=0}^{\infty} np(n) = E[X] \quad \text{and} \quad G(1) = E\left(1^X\right) = 1.$$

Note that

$$p(0) = G(0), \quad p(1) = G'(0), \quad p(2) = \frac{G''(0)}{2}, \ldots, p(n) = \frac{G^{(n)}(0)}{n!}$$

so that the p.d.f. can be recovered from the probability generating function, and if $G_X(z) = G_Y(z)$, then the p.d.f.'s for X and Y are the same.

We will find the probability generating function central to our computations for branching processes that are studied in Chapter 3. The following theorems will be used there.

Theorem 1.10

If $X_1, X_2, \ldots, X_n$ are independent random variables taking values in the nonnegative integers and $S_n = X_1 + X_2 + \cdots + X_n$, then

$$G_{S_n}(z) = \prod_{i=1}^{n} G_{X_i}(z).$$

Proof

We have that if $X_1, X_2, \ldots, X_n$ are independent random variables taking values in the nonnegative integer and ($S_n = X_1 + X_2 + \cdots + X_n$), then

$$E\left[z^{S_n}\right] = E\left[z^{X_1+X_2+\cdots+X_n}\right] = E\left[\prod_{i=1}^{n} z^{X_i}\right]$$

$$= \prod_{i=1}^{n} E\left[z^{X_i}\right] = \prod_{i=1}^{n} G_{X_i}(z)$$

where the second to last equality is valid because the X_i are independent.

Corollary

If $X_1, X_2, \ldots, X_n$ are independent, identically distributed random variables taking values in the nonnegative integers and $S_n = X_1 + X_2 + \cdots + X_n$, then

$$G_{S_n}(z) = [G_X(z)]^n$$

where X has the same distribution as each X_i.

Examples

(a) Let X be the Bernoulli random variable with $P(X = 1) = p$; $P(X = 0) = 1 - p$.
Then

$$G(z) = \sum p(x) z^x = (1-p)z^0 + pz^1 = (1-p) + pz.$$

(b) Let Y be the binomial random variable $Y = X_1 + \cdots + X_n$ where each X_i is the random variable of part (a).
Then

$$G_Y(z) = G_{X_1+\cdots+X_n}(z) = G_{X_1}(z) \cdots G_{X_n}(z) = \left[(1-p) + pz\right]^n.$$

(c) Let X be the Poisson random variable with

$$P(X = k) = e^{-\lambda}\frac{\lambda^k}{k!}, \quad k = 0, 1, 2, \ldots.$$

Then

$$G(z) = \sum_{k=0}^{\infty} e^{-\lambda}\frac{\lambda^k}{k!} z^k = e^{-\lambda}\sum_{k=0}^{\infty}\frac{\lambda^k}{k!} z^k = e^{-\lambda}e^{\lambda z} = e^{\lambda(z-1)}.$$

Theorem 1.11

Let $N, X_1, X_2, \ldots$ be independent random variables taking values in the nonnegative integers with $X_1, X_2, \ldots$ independent and identically distributed. Let $G_X(z)$ be the probability generating function of each X_i and $S_N = X_1 + X_2 + \cdots + X_N$. Then $G_{S_N}(z) = G_N(G_X(z))$.

Proof

We have

$$G_{S_N}(z) = E\left[z^{S_N}\right] = \sum_{n=0}^{\infty} E\left(z^{S_N} \mid N = n\right) P(N = n) = \sum_{n=0}^{\infty} E\left[z^{S_N}\right] P(N = n)$$

$$= \sum_{n=0}^{\infty} G_{X_1+X_2+\cdots+X_N}(z) P(N = n) = \sum_{n=0}^{\infty} [G_X(z)]^N P(N = n)$$

$$= G_N(G_X(z)).$$

In the second equality, we used the law of total probability.

Corollary

With the hypotheses of the previous theorem,

$$E[S_N] = E[N]E[X].$$

NOTE: This corollary will not be used for our study, but it is important in some probabilistic work and it fits with what we are doing now.

Proof

We have

$$\frac{d}{dz}\left[G_{S_N}(z)\right] = \frac{d}{dz}\left[G_N(G_X(z))\right].$$

Letting

$$u = G_X(z),$$

we have

$$\frac{d}{dz}\left[G_N(G_X(z))\right] = \frac{d}{du}[G_N(u)]\frac{d}{dz}[G_X(z)].$$

Now

$$\frac{d}{dz}G_X(z)\Big|_{z=1} = E[X]$$

and

$$\frac{d}{du}G_N(u)\Big|_{u=1} = E[N].$$

Set $z=1$. Then $u = G_X(1) = 1$, so

$$E[S_N] = \frac{d}{dz}G_{S_N}(z)\Big|_{z=1} = \frac{d}{dz}\left[G_N\left(G_X(z)\right)\right]\Big|_{z=1} = \frac{d}{du}[G_N(u)]\frac{d}{dz}[G_X(z)]\Big|_{z=1}$$

$$= \left(\frac{d}{du}[G_N(u)]\Big|_{u=1}\right)\left(\frac{d}{dz}[G_X(z)]\Big|_{z=1}\right) = E[N]E[X].$$

Moment Generating Functions

If the random variable X takes values in other than nonnegative integers, then the probability generating function is not as useful as the moment generating function. The moment generating function, $M_X(t)$ (or simply $M(t)$ if there is no possibility of confusing the random variables), is defined by

$$M_X(t) = E\left[e^{tX}\right] = \begin{cases} \sum_{n\in\mathbb{Z}} e^{tn}p(n) & \text{if } X \text{ is discrete} \\ \int_{-\infty}^{\infty} e^{tx}f(x)\,dx & \text{if } X \text{ is continuous} \end{cases}.$$

The most useful property of the moment generating function is

$$\frac{d^n}{dt^n}\left[M_X(t)\right]\Big|_{t=0} = E\left[X^n\right]$$

as can be seen from

$$\frac{d^n}{dt^n}E\left[e^{tX}\right]\Big|_{t=0} = E\left[\frac{d^n}{dt^n}e^{tX}\right]\Bigg|_{t=0} = E\left[X^n e^{tX}\right]\Big|_{t=0} = E\left[X^n\right]$$

where we assume that moving differentiation inside the summation (integral) is legitimate.

The strong law of large numbers and the central limit theorem are the two most important theorems in probability. We will have several uses for the former.

Theorem 1.12 (Strong Law of Large Numbers)

Let X_1, X_2, … be a sequence of independent, identically distributed random variables with $E[X_i] = \mu$. Then, with probability 1,

$$\lim_{n\to\infty} \frac{X_1 + X_2 + \cdots + X_n}{n} = \mu.$$

Exercises

1.1 For A and B events, show that the probability exactly one occurs is

$$P(A) + P(B) - 2P(A \cap B).$$

1.2 Let A_1, A_2, …, A_n be events. Let B_1, B_2, …, B_n be events defined by

$$B_1 = A_1, \quad B_2 = A_2 \backslash A_1, \ldots, B_k = A_k \backslash \left(\cup_{i=1}^{k-1} A_i. \right).$$

(a) Show that the sets B_1, B_2, …, B_n are mutually exclusive.

(b) Show that

$$P\left(\cup_{i=1}^{n} A_i \right) = \sum_{i=1}^{n} P(B_i).$$

1.3 (a) We roll one die 12 times. We count the number of times we roll a 3. What is the sample space?

(b) We roll 12 dice and count the number of 3s. What is the sample space?

1.4 We roll 2 dice and flip 2 coins. Describe the sample space. (It may help to consider the dice to be of different colors and the coins to be of different denominations.)

1.5 A baseball player hits a single with probability .2, a double with probability .05, a triple with probability .01, and a home run with probability .04.

(a) In a game where he has 5 at bats, what is the probability he has exactly 3 hits?

(b) If he has exactly 3 hits in the 5 at bats, what is the probability exactly 2 of those hits are singles?

1.6 Alice and Betty alternate shooting free throws, with Alice going first. Alice makes 50% of her free throws and Betty makes 70% of hers. What is the probability that Betty is the first to make a free throw?

1.7 If A and B are independent events,

(a) What can you say about the independence of A^c and B^c?

(b) What can you say about the independence of A^c and B?

1.8 There are two restaurants, A and B, in a town. If George eats at restaurant A, the next time he goes out, he eats at restaurant B with probability .75, and if he eats at restaurant B, the next time he goes out, he eats at restaurant A with probability .5. If he first goes to restaurant A with probability .4, what is the probability he will eat at restaurant B the third time he goes out?

1.9 Katie has to decide between taking a biology course or a chemistry course. She likes biology better, but the probability that she will get an A in the biology course is .4 and the probability that she will get an A in the chemistry course is .6. She decides to base her course on roll of a die. If the die comes up 1, 2, 3, or 4, she will take the biology course. What is the probability she will get an A?

1.10 In this exercise, we show that the negative binomial distribution is a probability distribution. Central to the derivation is the formula

$$(1-x)^{-r} = \sum_{k=0}^{\infty} \binom{k+r-1}{r-1} x^k.$$

We must show

$$\sum_{n=r}^{\infty} \binom{n-1}{r-1} (1-p)^{n-r} p^r = 1. \tag{1}$$

(a) In the left-hand side of Equation 1, set $k = n - r$ and show this converts that term to

$$\sum_{k=0}^{\infty}\binom{k+r-1}{r-1}(1-p)^k p^r.$$

(b) Use the expression for $(1 - x)^{-r}$ to show the sum is 1.

1.11 A family has 4 children. At least 1 is a boy. What is the probability that there are exactly 2 girls?

1.12 We have $f(x,y) = C(x + 2y)$, $x = 0, 1, 2, 3$; $y = 1,3$.

(a) Find C so that $f(x,y)$ is a joint p.d.f.

(b) Compute the marginal p.d.f.'s f_X and f_Y.

(c) Are X and Y independent?

(d) Compute $E[X^2 + XY]$.

1.13 In the following is a table for a bivariate distribution of random variables X and Y.

x	−1	0	2	4
y				
1	1/8	0	1/8	1/4
2	1/6	1/6	0	1/6

(a) Are X and Y independent?

(b) Find the marginal distributions of X and Y.

(c) Find $E[X]$, $E[XY]$, and $E[X^2]$.

1.14 Suppose that X_1 and X_2 are independent Poisson random variables with parameters λ_1 and λ_2, respectively. Find the p.d.f. of $X_1 + X_2$.

1.15 A family has two children. Assume that a child is equally likely to be a boy or a girl.

(a) What is the probability both will be boys if the oldest is a boy?

(b) What is the probability both will be boys if one is a boy?

1.16 This is a problem connected with the Ehrenfest model, which we will study in later chapters. We have 10 balls numbered 1 through 10 and 2 urns. The balls are placed at random into one of the two urns. We choose a number from 1 to 10 and move that ball from the urn it is in to the different urn.

(a) If urn 1 has 3 balls at the beginning of the experiment, what is the probability that it has 2 balls after the experiment?

(b) If urn 1 has 6 balls at the beginning of the experiment, what is the probability that it has 7 balls after the experiment?

1.17 We have a system that consists of three components. Suppose that at any given time the first component fails with probability .2, the second with probability .3, and the third with probability .1. If the components are independent of one another, what is the probability that the system fails if

(a) The system fails if any two of the components fail?

(b) The system fails if any one of the components fails?

1.18 In the game Trouble®, where one die is rolled, to move your token, you must first roll a 6. What is the expected number of rolls you must make before you can move your token?

1.19 (a) Give an example of a nonnegative random variable X for which

$$P(X<\infty)=1 \quad \text{and} \quad E[X]=\infty.$$

(b) Show that if X is a nonnegative random variable for which

$$P(X<\infty)<1 \quad \text{then } E[X]=\infty.$$

1.20 We are at a party where there are 10 people. Each person has worn a coat, and the coats are in a dark room. At the end of the party, each person randomly picks up a coat. Let X_i be the random variable defined by

$$X_i=\begin{cases}1 & \text{if person } i \text{ selects her own coat}\\ 0 & \text{otherwise}\end{cases}.$$

(a) Are the random variables independent?

(b) What is the expected value of X_i?

(c) What is the expected value of the number of people that will get their own coat?

1.21 Let X be a random variable with p.d.f.

$$p(1)=\frac{1}{4},\quad p(3)=\frac{1}{2},\quad p(4)=\frac{1}{8},\quad p(7)=\frac{1}{8}.$$

(a) Find the probability generating function for X.

(b) Use the probability generating function to find $E[X]$.

1.22 You ask a friend to be responsible for inviting guests to a party. You are particularly interested that your boss be there. Your friend will send out the invitations with probability .8, and if your boss receives an invitation, she will attend with probability .6.

(a) What is the probability that your boss will attend?

(b) If your boss does not attend, what is the probability that she received an invitation?

1.23 Prove that if X is a Poisson random variable with parameter λ, then

$$E\left[X^n\right] = \lambda E\left[(X+1)^{n-1}\right].$$

1.24 In Markov chains, an important characteristic for a random variable to have is that its p.d.f. be memoryless. We will show in Chapter 4 that the only continuous random variables that are memoryless are those with an exponential p.d.f. In this exercise, we show that geometric random variables (which are discrete) are memoryless. To do this, show that if X is a geometric random variable, then

$$P(X = n + k \mid X > n) = P(X = k).$$

In fact, the only memoryless discrete random variables are geometric random variables.

1.25 We roll two dice. Let X be the larger of the two values and Y be the sum of the two values. For the roll (1,1), $X = 1$, $Y = 2$.

(a) What is the sample space?

(b) Construct the joint p.d.f. of X and Y.

(c) Determine the marginal density functions for X and Y.

(d) Are X and Y independent?

(e) Compute $E[XY + Y]$.

1.26 We roll a red die and a green die. Let X be value of the red die and Y be the value of the green die.

(a) Construct the joint p.d.f. of X and Y.

(b) Determine the marginal density functions for X and Y.

(c) Are X and Y independent?

(d) Compute $E[XY + Y]$.

1.27 Roll two dice. If the dice show different numbers, what is the probability that one is a 6?

1.28 Flip a coin until a head comes up. The coin is biased so that the probability of getting a head is .6. If the first flip is a tail, what is the probability more than three flips will be required?

1.29 A series of random digits 0, 1, …, 9 are generated. How long must the series be before the probability of getting a 1 in the series is at least .8?

1.30 Customers arrive at a store at the rate of 8 per hour according to a Poisson distribution. What is the probability on a given hour that more than 6 customers will arrive?

1.31 Suppose X_1 and X_2 are independent Bernoulli random variables with $P(X_i = 1) = p$. Find the distribution of $X_1 + X_2$ using

(a) The probability generating functions

(b) Convolution

1.32 Suppose X_1 and X_2 are independent binomial random variables each with n trials and probability of success of a trial being p. Find the distribution of $X_1 + X_2$ using

(a) The probability generating functions

(b) Convolution

1.33 (a) Find the probability generating function for a geometric random variable where the probability of success on a given trial is p.

(b) Suppose X_1 and X_2 are independent geometric random variables, each probability of success of a trial being p. Find the distribution of $X_1 + X_2$.

1.34 This problem appeared in a Sunday magazine of a newspaper and garnered a lot of attention after the author gave the correct answer, and several mathematicians incorrectly disagreed with her. It has become known as the *Monty Hall problem* named after the emcee of a television game show.

There are three doors. Behind two of doors are booby prizes, and behind the other door is a prize of value. The contestant selects a door and will receive the prize behind that door. The emcee, knowing where the valuable prize is, opens one of the doors that the contestant does not select, revealing a booby prize. The contestant now has the opportunity to switch her choice of doors. The question is, does it affect her chances of winning and should she do so?

One approach to the problem is to use conditional probability, but that may not be the simplest.

1.35 This problem, which is famous in the literature, is an elementary version of an area known as hidden Markov models.

We have two coins in our pocket. One is a normal coin, with a head and a tail; the other coin has two heads. We select a coin and, not knowing which coin we have selected, flip it, and the outcome is a head.

(a) What is the probability we flipped the biased coin?

(b) We flip the coin a second time, and again the outcome is a head. Now what is the probability we are flipping the biased coin?

A common way to introduce hidden Markov models is with the *dishonest casino problem*. In this problem, a casino has two dice: one fair and the other biased with some number, say, 6, being more likely than the other numbers. The croupier can switch the dice, but doesn't do so very often because of fear of detection. The problem is, given a long string of rolls of the dice, determine during which parts of the string were the different dice used.

This turns out to have several applications. One application is in biology, where it can be important in determining at which locations on DNA are genes most likely to reside. DNA consists of four nucleotides named A, C, G, and T that appear in random order. In certain *islands* of a strand of DNA, the C and G nucleotides appear more often than on the rest of the strand, and it has been shown that these islands are where genes are most likely located. If the islands can be located, it greatly simplifies the problem of locating the genes.

1.36 Let X_i be independent random variables with

$$P(X_i = 1) = .4, \quad P(X_i = 3) = .6$$

and let N be the random variable with

$$P(N = n) = \frac{1}{2^n}, \quad n = 1, 2, \ldots.$$

Find $G_{S_N}(z)$.

1.37 Show that if X and Y are independent and identically distributed random variables that take values in the nonnegative integers with common density distribution f and cumulative distribution F, then $X + Y$ has the cumulative distribution

$$(F * f)(x) = \sum_{y=0}^{x} F(n - y) f(y).$$

2

Discrete-Time, Finite-State Markov Chains

Introduction

In this chapter, our setting is we have a system that consists of a finite number of states $s_1, \ldots, s_n$ and a process that is in exactly one of these states at any given time. For a discrete-time Markov chain, one can visualize the process running according to a digital clock and the process making the decision of how to change states at each change in the clock. In making the change of states, the process is governed by a set of conditional probabilities, called transition probabilities. These are defined by

$P_{i,j}(n)$ = the probability that the process changes from state s_i to s_j when the time changes from n to $n + 1$

$P_{i,i}(n)$ = the probability that the process remains in state s_i when the time changes from n to $n + 1$.

As we have defined these probabilities, there is a dependency on n. However, we will assume that the process is homogeneous in time so that

$$P_{i,j}(n) = P_{i,j}(m) \text{ for all nonnegative integers } m \text{ and } n$$

and $P_{i,j}$ will denote the transition probability from state s_i to state s_j at any time.

Notation

To make the notation less cumbersome, we will denote state s_i by i.

Intuitively, the Markov property, which we define below, says that what the process does in the next step depends only on where it is at the present time and its prior history is of no consequence.

To express the Markov property in the discrete case mathematically, suppose that N is a countable index set thought of as time (usually the positive integers or nonnegative integers) and that S is the state space. For each $n \in N$, X_n is a random variable that gives the state of the process at time n. The Markov property is

$$P\left(X_n = i_n \middle| X_{n-1} = i_{n-1}, \ldots, X_1 = i_1, X_0 = i_0\right) = P\left(X_n = i_n \middle| X_{n-1} = i_{n-1}\right)$$

for any for any $i_0, i_1, \ldots, i_{n-1}, i_n \in S$.

There is an equivalent condition that appears more general which we note in the first theorem.

Theorem 2.1

Let S denote the state space of a Markov chain. Then for any

$$s, i_0, i_1, \ldots, i_{n-1} \in S \text{ and integers } n \geq 1, m \geq 0$$

we have

$$P\left(X_{n+m} = s \middle| X_{n-1} = i_{n-1}, \ldots, X_1 = i_1, X_0 = i_0\right) = P\left(X_{n+m} = s \middle| X_{n-1} = i_{n-1}\right).$$

The proof is left as an exercise.

The theorem says that for a Markov process, conditioning a future event on a given set of previous outcomes is equivalent to conditioning only on the most recent of the outcomes in the set.

Transition Matrices

The major tool in doing computations for finite, discrete-time Markov chains is the transition matrix, which is a matrix whose entries are the transition probabilities. This matrix is typically denoted by P. Thus,

$$P = \begin{pmatrix} P_{1,1} & \cdots & P_{1,n} \\ \vdots & & \vdots \\ P_{n,1} & \cdots & P_{n,n} \end{pmatrix},$$

where $P_{i,j}$ is the transition probability from state i to state j.

Note that $P_{i,j} \geq 0$ and

$$\sum_{j=1}^{n} P_{i,j} = 1, \quad \text{for all } i = 1, \ldots, n.$$

Matrices that have these two properties are called stochastic matrices.

It is not necessarily true that $P_{i,j} = P_{j,i}$ nor is it necessarily true that

$$\sum_{i=1}^{n} P_{i,j} = 1, \quad \text{for all } j = 1, \ldots, n.$$

The transition matrix is useful in computations for several reasons, the most prominent being that the probability in going from state i to state j in n steps is given by

$$\left(P^n\right)_{i,j}$$

as we now explain.

In the case $n = 2$, if we go from state i to state j in 2 steps, then on the first step we go from state i to some state k, and on the second step, we go from state k to state j. For a given state k, the probability of this is

$$P_{i,k}P_{k,j}.$$

To determine the probability of going from state i to state j in 2 steps, we sum over all values of k to get

$$P_{i,1}P_{1,j} + P_{i,2}P_{2,j} + \cdots P_{i,n}P_{n,j} = \sum_{k=1}^{n} P_{i,k}P_{k,j} = \left(P^2\right)_{i,j}.$$

The extension to all values of n can be accomplished by induction.

Notation: We will also use the notation $p_n(i, j)$ for $(P^n)_{i,j}$.

The Chapman–Kolmogorov Equation, given in the next theorem, is the mathematical statement of the following intuitive idea: to go from state i to state j in $n + m$ steps, we go from i to some state k in n steps and then from k to j in m steps. If we first fix k, the probability of this occurring is

$$P\left(X_n = k \mid X_0 = i\right)P\left(X_{n+m} = j \mid X_n = k\right).$$

Now let state k vary. If we sum over all states k, the probability of going from state i to state j in $n + m$ steps is

$$\sum_{k} P\left(X_n = k \mid X_0 = i\right)P\left(X_{n+m} = j \mid X_n = k\right).$$

Because of time homogeneity, we have

$$P(X_{n+m} = j \mid X_n = k) = P(X_m = j \mid X_0 = k),$$

so

$$\sum_k P(X_n = k \mid X_0 = i) P(X_{n+m} = j \mid X_n = k)$$
$$= \sum_k P(X_n = k \mid X_0 = i) P(X_m = j \mid X_0 = k).$$

We formalize this idea.

Theorem 2.2 (Chapman–Kolmogorov Equation)

For m and n nonnegative integers

$$p_{n+m}(i,j) = \sum_k p_m(i,k) p_n(k,j).$$

Proof
We have

$$P_{n+m}(i,j) = P(X_{n+m} = j \mid X_0 = i)$$
$$= \sum_{k \epsilon S} P(X_{n+m} = j, X_m = k \mid X_0 = i) = \sum_{k \epsilon S} \frac{P(X_{n+m} = j, X_m = k, X_0 = i)}{P(X_0 = i)}$$
$$= \sum_{k \epsilon S} \frac{P(X_{n+m} = j, X_m = k, X_0 = i)}{P(X_m = k, X_0 = i)} \frac{P(X_m = k, X_0 = i)}{P(X_0 = i)}.$$

Now,

$$\frac{P(X_{n+m} = j, X_m = k, X_0 = i)}{P(X_m = k, X_0 = i)} = P(X_{n+m} = j \mid X_m = k, X_0 = i).$$

By the Markov property,

$$P(X_{n+m} = j \mid X_m = k, X_0 = i) = P(X_{n+m} = j \mid X_m = k)$$

and by time homogeneity

$$P(X_{n+m} = j \mid X_m = k) = P(X_n = j \mid X_0 = k).$$

Thus,

$$\sum_{k \epsilon S} \frac{P\left(X_{n+m}=j, X_m=k, X_0=i\right)}{P\left(X_m=k, X_0=i\right)} \frac{P\left(X_m=k, X_0=i\right)}{P\left(X_0=i\right)}$$

$$=\sum_{k \epsilon S} P\left(X_{n+m}=j \mid X_m=k\right) P\left(X_m=k \mid X_0=i\right)$$

$$=\sum_{k \epsilon S} P\left(X_n=j \mid X_0=k\right) P\left(X_m=k \mid X_0=i\right)$$

$$=\sum_{k \epsilon S} p_m(i, k) p_n(k, j).$$

Theorem 2.3

If A and B are $n \times n$ stochastic matrices, then AB is an $n \times n$ stochastic matrix.

Proof

Let $C = AB$.

We know $A_{ij} \geq 0$ and $B_{ij} \geq 0$, so $C_{ij} \geq 0$. We also know

$$\sum_{k=1}^{n} A_{ik} = 1 \text{ for every } i \quad \text{and} \quad \sum_{j=1}^{n} B_{kj} = 1 \text{ for every } k.$$

We must show

$$\sum_{j=1}^{n} C_{ij} = 1 \text{ for every } i.$$

We have

$$\sum_{j=1}^{n} C_{ij} = \sum_{j=1}^{n}\left(\sum_{k=1}^{n} A_{ik} B_{kj}\right) = \sum_{k=1}^{n}\left(\sum_{j=1}^{n} A_{ik} B_{kj}\right)$$

$$= \sum_{k=1}^{n} A_{ik}\left(\sum_{j=1}^{n} B_{kj}\right) = \sum_{k=1}^{n} A_{ik}(1) = 1.$$

Corollary

If P is a transition matrix, then P^n is a transition matrix.

Definition

A vector

$$\hat{v} = (v_1, \ldots, v_n) \quad \text{or} \quad \hat{v} = \begin{pmatrix} v_1 \\ \vdots \\ v_n \end{pmatrix}$$

is a probability vector if if $v_i \geq 0$ and $v_1 + \cdots + v_n = 1$.

The purpose of the next example is to develop some intuition for how we use the transition matrix to model the evolution of a process.

Example

Suppose we have three containers and we divide 1000 balls among the three containers. These three containers represent the states of the system. We initially place 500 balls in the first container, 300 in the second, and 200 in the third. So, 50% of the balls are in the first container (state 1), 30% are in the second container (state 2), and 20% are in the third (state 3). So for any given ball, the probability of that ball beginning in state 1 is .5, the probability of that ball beginning in state 2 is .3, and the probability of that ball beginning in state 3 is .2.

When an alarm sounds, 1/4 of the balls in the first container stay in the first container (so that the probability of remaining in state 1 is 1/4), 1/2 of the balls in the first container move to the second container, and 1/4 of the balls in the first container move to the third container. In terms of the transition matrix, we have said

$$P_{11} = \frac{1}{4}, \quad P_{12} = \frac{1}{2}, \quad P_{13} = \frac{1}{4}.$$

Also, suppose that none of the balls in the second container move to the first, 1/2 stay in the second, and 1/2 move to the third. Thus,

$$P_{21} = 0, \quad P_{22} = \frac{1}{2}, \quad P_{23} = \frac{1}{2}.$$

For the third container, 1/8 of the balls move to the first container, 3/4 to the second, and 1/8 stay in the third. Thus,

$$P_{31} = \frac{1}{8}, \quad P_{32} = \frac{3}{4}, \quad P_{33} = \frac{1}{8}.$$

So the transition matrix is

$$P = \begin{pmatrix} \frac{1}{4} & \frac{1}{2} & \frac{1}{4} \\ 0 & \frac{1}{2} & \frac{1}{2} \\ \frac{1}{8} & \frac{3}{4} & \frac{1}{8} \end{pmatrix}.$$

We now compute how many balls are in each container after the transition.
The first container will have

$$(500)\frac{1}{4} \text{ (the number that remained in the first container)}$$
$$+(300)0 \text{ (the number that moved from the second container to the first)}$$
$$+(200)\frac{1}{8} \text{ (the number that moved from the third to the first).}$$

So the first container will have

$$(500)\frac{1}{4}+(300)0+(200)\frac{1}{8}=150$$

balls after the transition.
The second container will have

$$(500)\frac{1}{2} \text{ (the number that moved from the first to the second)}$$
$$+(300)\frac{1}{2} \text{ (the number that remained in the second container)}$$
$$+(200)\frac{3}{4} \text{ (the number that moved from the third to the second).}$$

So, the second container will have

$$(500)\frac{1}{2}+(300)\frac{1}{2}+(200)\frac{3}{4}=550$$

balls after the transition.

The third container will have

$$(500)\frac{1}{4}(\text{the number that moved from the first to the third})$$

$$+(300)\frac{1}{2}\ (\text{the number that moved from the second to the third})$$

$$+(200)\frac{1}{8}(\text{the number that remained in the third container}).$$

So, the third container will have

$$(500)\frac{1}{4}+(300)\frac{1}{2}+(200)\frac{1}{8}=300$$

balls after the transition.

If we take $\hat{v}=(500,300,200)$, then

$$\hat{v}P=(500,300,200)\begin{pmatrix} \frac{1}{4} & \frac{1}{2} & \frac{1}{4} \\ 0 & \frac{1}{2} & \frac{1}{2} \\ \frac{1}{8} & \frac{3}{4} & \frac{1}{8} \end{pmatrix}$$

$$=\left((500)\frac{1}{4}+(300)0+(200)\frac{1}{8},(500)\frac{1}{2}+(300)\frac{1}{2}+(200)\frac{3}{4},(500)\frac{1}{4}+(300)\frac{1}{2}+(200)\frac{1}{8}\right)$$

$$=(150,550,300).$$

Thus, if $\hat{v}$ is the present distribution of the system, then $\hat{v}P$ is the distribution of the system one step later.

In our later work, the *input vector* $\hat{v}$ will give the probability that the system begins in various states.

Theorem 2.4

If P is a transition matrix and

$$\hat{v}=(v_1,\ldots,v_n)$$

is a probability vector, then $\hat{v}P$ is a probability vector.

Proof

We have $\hat{v}P$ is an n-vector whose entries are nonnegative. We must show the sum of the entries is 1. Now

$$\hat{v}P = (v_1, \ldots, v_n)\begin{pmatrix} p_{11} & \cdots & p_{1n} \\ \vdots & & \vdots \\ p_{n1} & \cdots & p_{nn} \end{pmatrix} = \begin{pmatrix} p_{11}v_1 + \cdots + p_{n1}v_n \\ \vdots \\ p_{1n}v_1 + \cdots + p_{nn}v_n \end{pmatrix} = \begin{pmatrix} \sum_{i=1}^{n} p_{i1}v_i \\ \vdots \\ \sum_{i=1}^{n} p_{in}v_i \end{pmatrix},$$

so

$$\sum_{j=1}^{n} (\hat{v}P)_j = \sum_{j=1}^{n}\left(\sum_{i=1}^{n} p_{ij}v_i\right) = \sum_{i=1}^{n}\left(\sum_{j=1}^{n} p_{ij}v_i\right) = \sum_{i=1}^{n} v_i\left(\sum_{j=1}^{n} p_{ij}\right)$$

$$= \sum_{i=1}^{n} v_i(1) = 1.$$

Repeating an earlier idea for emphasis: if $\hat{v}$ is the probability vector whose ith entry is the probability the system is in state i at time n and P is the transition matrix for the process, then $\hat{v}P$ is the probability vector whose ith entry is the probability the system is in state i at time $n + 1$.

The next theorem will be important as we characterize the evolution of a Markov process.

Theorem 2.5

A stochastic matrix has an eigenvalue of 1 and no eigenvalue greater than 1.

Proof

Recall that the eigenvalues of a square matrix P and its transpose, P^T, are the same, although the eigenvector $\hat{v}$ for P with the eigenvalue λ will not necessarily be the transpose of the eigenvector for P^T that has eigenvalue λ.

Note that if P is a stochastic matrix, then

$$P\begin{pmatrix} 1 \\ 1 \\ \vdots \\ 1 \end{pmatrix} = \begin{pmatrix} 1 \\ 1 \\ \vdots \\ 1 \end{pmatrix},$$

so 1 is an eigenvalue of P.

Suppose that $\lambda > 1$ is an eigenvalue of P with eigenvector

$$\hat{x} = \begin{pmatrix} x_1 \\ x_2 \\ \vdots \\ x_N \end{pmatrix}$$

and suppose that $x_k = \max\{x_1, x_2, \ldots, x_N\}$. The kth coordinate of $P\hat{x}$ is

$$\begin{aligned} p_{k1}x_1 + p_{k2}x_2 + \cdots + p_{kN}x_N &\le p_{k1}x_k + p_{k2}x_k + \cdots + p_{kN}x_k \\ &= (p_{k1} + p_{k2} + \cdots + p_{kN})x_k \\ &= 1x_k < \lambda x_k. \end{aligned}$$

Thus, $\lambda > 1$ cannot be an eigenvalue of P.

Definition

A matrix P is a positive matrix if $P_{ij} > 0$ for all i, j. A matrix P is a nonnegative matrix if $P_{ij} \ge 0$ for all i, j.

If P is a positive transition matrix for a Markov chain, then it is possible to go from any state i to any state j in one step.

We will see that finite Markov chains that have positive transition matrices are particularly nice in that all initial probability distributions will evolve over a long period of time to the same probability distribution, which we call the equilibrium state of the process.

The next result will be useful in our discussion of particular types of Markov chains called absorbing Markov chains.

Theorem 2.6

Suppose that A is an $N \times N$ nonnegative matrix and the sum of the entries of each row is less than 1. Then the maximum eigenvalue of A is less than 1.

Proof

Suppose that λ is an eigenvector of A and $\hat{x}A = \lambda\hat{x}$. Then,

$$\begin{aligned} \hat{x}A &= (x_1, \ldots, x_n)\begin{pmatrix} a_{11} & \cdots & a_{1N} \\ \vdots & & \vdots \\ a_{N1} & \cdots & a_{NN} \end{pmatrix} \\ &= (a_{11}x_1 + \cdots + a_{N1}x_N, \ldots, a_{1N}x_1 + \cdots + a_{NN}x_N). \end{aligned}$$

Summing the entries of $\hat{x}A$ gives

$$(a_{11} + \cdots + a_{1N})x_1 + \cdots + (a_{N1} + \cdots + a_{NN})x_N < x_1 + \cdots + x_N.$$

The sum of the entries of $\lambda\hat{x}$ is

$$\lambda x_1 + \cdots + \lambda x_N,$$

so the sum of the entries of $\lambda\hat{x}$ is the sum of the entries of $A\hat{x}$ only if $\lambda < 1$.

Directed Graphs: Examples of Markov Chains

Associated with each Markov chain is a directed graph (or digraph). Each state of the Markov chain corresponds to a vertex of the digraph, and if i and j are states of the Markov chain with $P(i, j) > 0$, there is a directed edge from the vertex associated with state i to the vertex associated with state j. Digraphs can sometimes give a more visual image of the transition possibilities than the transition matrix.

Example

Suppose a Markov chain has the transition matrix

$$P = \begin{pmatrix} 0 & .1 & 0 & .2 & .3 & .4 \\ .3 & .2 & 0 & .3 & .1 & 0 \\ .4 & 0 & .6 & 0 & 0 & 0 \\ 0 & .8 & 0 & .2 & 0 & 0 \\ 1 & 0 & 0 & 0 & 0 & 0 \\ 0 & 0 & 0 & 0 & 0 & 1 \end{pmatrix}.$$

We denote the states by {1,2,3,4,5,6}. The associated digraph is shown in Figure 2.1.

We next consider some of the most common examples of finite Markov chains.

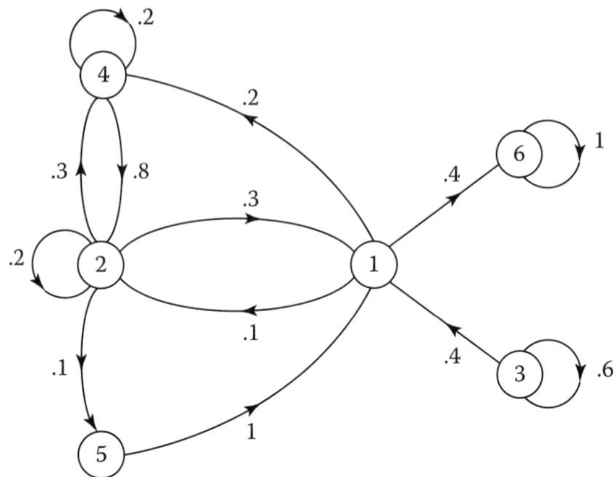

FIGURE 2.1
Digraph for Markov chain.

Random Walk with Reflecting Boundaries

A walker travels down a one-dimensional street that has a wall at each end. Suppose that it takes N steps to travel the length of the street. (See the digraph in Figure 2.2.) We keep track of the walker's position. We could do this by making the states of the Markov chain $\{0, 1, \ldots, N\}$, where if he is in state j it means he is j steps from the first wall. We suppose he takes a step forward from state k to state $k + 1$ with probability p if $k = 1, \ldots, N - 1$ and takes a step backward from state $k + 1$ to state k with probability $1 - p$ if $k = 1, \ldots, N - 1$.

We also suppose that he bounces off a wall when he is at state 0 or state N. Thus, if he is at state 0, he must move to state 1, and if he is at state N, he must move to state $N - 1$, so $P(0, 1) = 1$ and $P(N, N - 1) = 1$. The transition matrix for this process is

$$P = \begin{pmatrix} 0 & 1 & 0 & 0 & 0 & \cdots & 0 \\ 1-p & 0 & p & 0 & 0 & \cdots & 0 \\ 0 & 1-p & 0 & p & 0 & \cdots & 0 \\ \vdots & \vdots\ \vdots & & & \vdots & \vdots & \vdots \\ 0 & \cdots\ 0 & & & 1-p & 0 & p \\ 0 & \cdots\ 0 & & & 0 & 1 & 0 \end{pmatrix},$$

and the digraph is given in Figure 2.2.

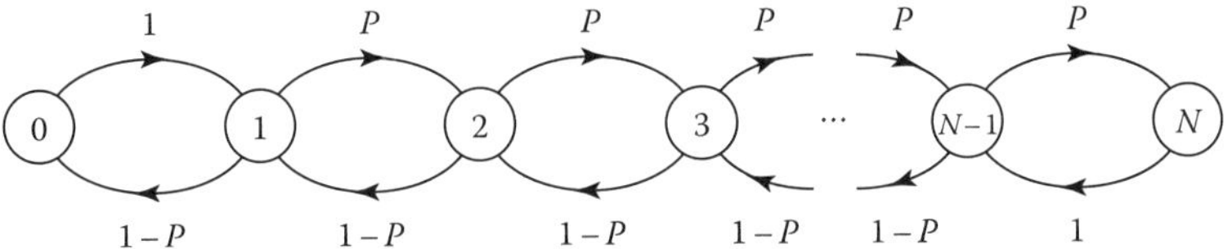

FIGURE 2.2
Digraph for Markov chain with absorbing boundaries.

Gambler's Ruin

Anne and Brieanne are playing a game of chance. At each play of the game, they bet \$1. Between them they have \$$N$. For each play of the game, Anne wins with probability p and Brieanne wins with probability $1 - p$. We want to keep track of how much each person has after n plays of the game. One way to do this is to let the states be Anne's fortune. Thus, the states of the chain are $\{0, 1, \ldots, N\}$.

The game must end when a player loses all her money; that is, if state is 0 or N. Once entering these states, we never leave, so $P(0, 0) = P(N, N) = 1$. Such states are called absorbing states. The transition matrix is

$$P = \begin{pmatrix} 1 & 0 & 0 & 0 & 0 & \cdots & 0 \\ 1-p & 0 & p & 0 & 0 & \cdots & 0 \\ 0 & 1-p & 0 & p & 0 & \cdots & 0 \\ \vdots & \vdots\ \vdots & & & \vdots & \vdots & \vdots \\ 0 & \cdots\ 0 & & & 1-p & 0 & p \\ 0 & \cdots\ 0 & & & 0 & 0 & 1 \end{pmatrix},$$

and the digraph is given in Figure 2.3.

The transition matrices of the two examples are similar, but their long-term behaviors are very different. The first process continues indefinitely, and it seems plausible that the second process must end. An important question for each process is what happens in the long run. For the wandering walker, this might be phrased what is the expected fraction of time that he

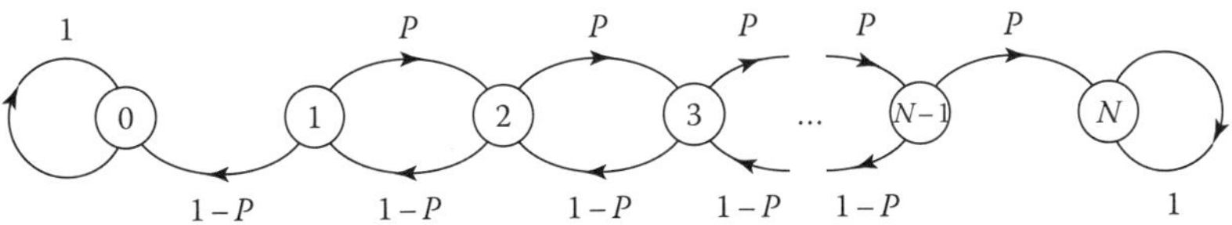

FIGURE 2.3
Digraph for Markov chain of gambler's ruin.

is in a particular state, and does this depend on the value of p or where he starts his journey.

For the gambler's ruin problem, the most pertinent questions deal with who is the more likely winner, and this is almost certainly affected by the value of p and the initial fortunes of the players.

Ehrenfest Model

The Ehrenfest model was developed by Paul Ehrenfest as a tool to understand the equilibrium of a gas. It is easier to visualize if we consider it phrased in terms of balls in urns.

We have M balls distributed between two urns that we denote urns 1 and 2. A ball is chosen at random and moved to the other urn. So each ball has a probability of $1/M$ of being chosen and this is independent of which urn contains the ball. (One could think of each ball having a different name, and to choose the ball we select a name not knowing which urn contains which names.) To describe the distribution of the balls, we let our states be the number of balls in urn 1. So the states are $\{0, 1, \ldots, M\}$. Then,

$$P(0,1)=1, \quad P(M,M-1)=1.$$

For $0 < m < M$, in order to go from state m to state $m - 1$, we must choose a ball from urn 1. If there are m balls in urn 1, then

$$P(m,m-1)=\frac{m}{M}.$$

In order to go from state m to state $m + 1$, we must choose a ball from urn 2. If there are m balls in urn 1, then there are $M - m$ balls in urn 2, so

$$P(m,m+1)=\frac{M-m}{M}.$$

The transition matrix in the case $M = 4$ is

$$P=\begin{pmatrix} 0 & 1 & 0 & 0 & 0 \\ 1/4 & 0 & 3/4 & 0 & 0 \\ 0 & 1/2 & 0 & 1/2 & 0 \\ 0 & 0 & 3/4 & 0 & 1/4 \\ 0 & 0 & 0 & 1 & 0 \end{pmatrix}.$$

We will investigate this model further in Exercise 2.17.

Central Problem of Markov Chains

A major goal of studying Markov chains is to determine the long-range behavior of the system. That is, given the transition matrix P and an initial probability vector $\hat{v}$, we would like to know whether

$$\lim_{n\to\infty} \hat{v}\,P^n$$

exists, and if so, what is the limit? Also, if the limit does exist, does it depend on $\hat{v}$? We first investigate the case where the number of states is two in order to develop some intuition.

In the case

$$P = \begin{pmatrix} 0 & 1 \\ 1 & 0 \end{pmatrix}$$

$$P^n = \begin{pmatrix} 0 & 1 \\ 1 & 0 \end{pmatrix} \text{ if } n \text{ is odd} \quad \text{and} \quad P^n = \begin{pmatrix} 1 & 0 \\ 0 & 1 \end{pmatrix} \text{ if } n \text{ is even.}$$

Thus, if $\hat{v} = (a, 1-a)$, then

$$\lim_{n\to\infty} \hat{v}P^{2n+1} = (a, 1-a)\begin{pmatrix} 0 & 1 \\ 1 & 0 \end{pmatrix} = (1-a, a)$$

and

$$\lim_{n\to\infty} \hat{v}\,P^{2n} = (a, 1-a)\begin{pmatrix} 1 & 0 \\ 0 & 1 \end{pmatrix} = (a, 1-a)$$

and the limit $\lim_{n\to\infty} \hat{v}\,P^n$ will exist if and only if $a = 1/2$ and $\hat{v} = (1/2, 1/2)$.

Next we consider

$$P = \begin{pmatrix} a & 1-a \\ b & 1-b \end{pmatrix} \quad 0 < a, b < 1.$$

If we use a computer algebra system (CAS) and compute P^n for several values of n, no obvious pattern appears. However, if we let $a = b$, then

$$P = \begin{pmatrix} a & 1-a \\ a & 1-a \end{pmatrix} \quad \text{and} \quad P^n = \begin{pmatrix} a & 1-a \\ a & 1-a \end{pmatrix}$$

and for any probability vector $(c, 1 - c)$

$$\lim_{n\to\infty}(c,1-c)P^n = (c,1-c)\begin{pmatrix} a & 1-a \\ a & 1-a \end{pmatrix} = (a,1-a).$$

This is the best of all possible worlds as far as long-term behavior is concerned. For any probability vector $\hat{v}$, we have that

$$\lim_{n\to\infty} \hat{v}P^n$$

exists and is always the same, regardless of what $\hat{v}$ is.

This example seems to be terribly restrictive on the form of P, and it is. The key to the behavior in the second example is that the rows of P are identical. We now embark on a project to find a condition on P that will ensure every row of P^n is the same for n large; that is,

$$\lim_{n\to\infty} P^n = \begin{pmatrix} a_1 & a_2 & \cdots & a_N \\ a_1 & a_2 & \cdots & a_N \\ \vdots & \vdots & \vdots & \vdots \\ a_1 & a_2 & \cdots & a_N \end{pmatrix}.$$

If this is the case, then we show in a corollary to the next theorem that for any probability vector $\hat{v}$

$$\lim_{n\to\infty} \hat{v}\, P^n = (a_1, a_2, \ldots, a_N).$$

NOTE: In the first part of this example, when we took $a = b$, it may well have appeared that we were taking a lucky guess. In fact, one can show this must be the case using eigenvectors (see Lawler 2006). We will see the underlying reason for this when we study consequences of the Perron–Frobenius Theorem.

Condition to Ensure a Unique Equilibrium State

Definition

A probability vector $\hat{\pi}$ for which $\hat{\pi}P = \hat{\pi}$ is called an equilibrium distribution (invariant distribution, stationary distribution, steady-state distribution) for the transition matrix P.

NOTE: Recall the example

$$P = \begin{pmatrix} 0 & 1 \\ 1 & 0 \end{pmatrix},$$

we found that $\hat{v} = (1/2, 1/2)$ was a steady state for P, but that if we chose any other initial probability vector $\hat{v}$, then

$$\lim_{n\to\infty} \hat{v} P^n$$

did not exist. This is analogous to a situation in dynamical systems where one has an unstable equilibrium state. We would like to have a condition on P so that for every initial probability vector $\hat{v}$

$$\lim_{n\to\infty} \hat{v} P^n$$

exists and has the same value.

As we saw above, some stochastic matrices P have the property that they have a unique equilibrium state and for every initial probability vector $\hat{v}$,

$$\lim_{n\to\infty} \hat{v} P^n$$

exists and has the same value which is that unique equilibrium state.

The next theorem and its corollary give a sufficient condition to determine such matrices.

Theorem 2.7

Let P be an $n \times n$ stochastic matrix such that $P_{ij} \geq \epsilon > 0$ for $1 \leq i, j \leq n$.

Suppose P_j is the jth column of P and that m is the minimum entry of P_j and M is the maximum entry of P_j. We assume $m < M$. Let m^* denote the minimum entry of the jth column of P^2 and M^* the maximum entry of the jth column of P^2. Then

$$M^* - m^* \leq (1 - 2\epsilon)(M - m).$$

NOTE: Any positive $n \times n$ matrix P has the property that there is an $\epsilon > 0$ for which $P_{ij} \geq \epsilon > 0$ for $1 \leq i, j \leq n$. (Just take ϵ to be the minimum of the entries of P.) With the proof of this theorem, we know that the entries of the columns of P^n become the same as $n \to \infty$.

Proof
Let

$$P = \begin{pmatrix} p_{11} & p_{12} & \cdots & p_{1n} \\ p_{21} & p_{22} & \cdots & p_{2n} \\ \vdots & \vdots & \vdots & \vdots \\ p_{n1} & p_{n2} & \cdots & p_{nn} \end{pmatrix}.$$

Choose a column of P, call it P_j, and for specificity, suppose the smallest entry of the column is m, the largest is M. To make the explanation more clear, we suppose the column is

$$\begin{pmatrix} m \\ x_2 \\ x_3 \\ \vdots \\ x_{n-1} \\ M \end{pmatrix}.$$

We show that each entry of the jth column of P^2 — that is, $(P^2)_j$ — is larger than m and smaller than M. (We are assuming $m < M$, which is the case unless all the entries of P_j are the same.)

The kth entry of $(P^2)_j$ is

$$p_{k1}m + p_{k2}x_2 + \cdots p_{k(n-1)}x_{n-1} + p_{kn}M.$$

We will show

$$m < p_{k1}m + p_{k2}x_2 + \cdots p_{k(n-1)}x_{n-1} + p_{kn}M < M.$$

Of course, we will not have strict inequality if $m = M$.

Now

$$\begin{aligned} p_{k1}m + p_{k2}x_2 + \cdots p_{k(n-1)}x_{n-1} + p_{kn}M &\geq p_{k1}m + p_{k2}m + \cdots p_{k(n-1)}m + p_{kn}M \\ &= p_{k1}m + p_{k2}m + \cdots p_{k(n-1)}m + p_{kn}\left[m + (M-m)\right] \\ &= p_{k1}m + p_{k2}m + \cdots p_{k(n-1)}m + p_{kn}m + p_{kn}(M-m) \\ &= \left(p_{k1} + p_{k2} + \cdots + p_{k(n-1)} + p_{kn}\right)m + p_{kn}(M-m) \\ &= m + p_{kn}(M-m) \geq m + \epsilon(M-m) \end{aligned}$$

since every $p_{ij} \geq \epsilon$.

Similarly,

$$\begin{aligned}
p_{k1}m + p_{k2}x_2 + \cdots p_{k(n-1)}x_{n-1} + p_{kn}M &\le p_{k1}m + p_{k2}M + \cdots p_{k(n-1)}M + p_{kn}M \\
&= p_{k1}\left[M + (m - M)\right] + p_{k2}M + \cdots p_{k(n-1)}M + p_{kn}M \\
&= p_{k1}M + p_{k2}M + \cdots p_{k(n-1)}M + p_{kn}M + p_{k1}(m - M) \\
&= (p_{k1} + p_{k2} + \cdots + p_{kn})M + p_{k1}(m - M) \\
&= M + p_{k1}(m - M) \le M + \epsilon(m - M)p \\
&= M - \epsilon(M - m).
\end{aligned}$$

Thus, any entry of the jth column of P^2, $p_{k1}m + p_{k2}x_2 + \cdots + p_{kn}M$, satisfies

$$m + \epsilon(M - m) \le p_{k1}m + p_{k2}x_2 + \cdots + p_{kn}M \le M - \epsilon(M - m),$$

so if α and β are any two entries of the kth column of P^2, we have

$$\begin{aligned}
|\alpha - \beta| &\le \left[M - \epsilon(M - m)\right] - \left[m + \epsilon(M - m)\right] = (M - m) - 2\epsilon(M - m) \\
&= (1 - 2\epsilon)(M - m).
\end{aligned}$$

Corollary

Let P be a matrix as in the theorem. Then $\lim_{n\to\infty}\left(P^n\right)_j$ is a vector whose components are all the same.

Proof

Let M_n be the maximum entry of the jth column of P^n. We showed in the proof of the theorem that $\{M_n\}$ is a decreasing sequence and it is bounded below by 0. Thus, $\{M_n\}$ converges, say to $\bar{M}$.

Similarly, the sequence of the minimum entries of the jth column of P^n, $\{m_n\}$, is a bounded increasing sequence that converges, say to $\bar{m}$.

Since $m_n \le M_n$ for every n, we have $\bar{m} \le \bar{M}$, but if $\bar{m} < \bar{M}$, we can apply the theorem to the limit matrix to reach a contradiction.

Example

Consider the transition matrix

$$P = \begin{pmatrix} .9 & .1 \\ .3 & .7 \end{pmatrix}.$$

In the theorem, ϵ is the smallest entry of P which in this case is .1. In the first column $M = .9$ and $m = .3$ so $M - m = .6$. The theorem asserts that if we raise P to the 2nd power, the difference between the largest and the smallest entries of the first column P^2 will be less than

$$(1-2\epsilon)(M-m) = .8 \times .6.$$

In this case,

$$P^2 = \begin{pmatrix} .84 & .16 \\ .48 & .52 \end{pmatrix}.$$

We see that the difference between largest and the smallest entries of the first column of P^2 is

$$.84 - .48 = .36$$

and

$$.8 \times .6 = .48.$$

For the second column, $M = .7$ and $m = .1$, so $M - m = .6$.
and the difference between the largest and smallest entries of the second column of P^2 is

$$.52 - .16 = .36,$$

which is also smaller than

$$.8 \times .6 = .48.$$

Corollary

Suppose that P is a positive stochastic matrix. Then

$$\lim_{n\to\infty} P^n = A,$$

where A is a stochastic matrix with positive entries, and the rows of A are identical.

Proof

The proof is the observation that if the entries of each column of a matrix are the same, then the rows of the matrix are the same.

NOTE: By the statement

$$\lim_{n\to\infty} P^n = A$$

we mean given $\epsilon > 0$, there is a number $N(\epsilon)$ such that if $n > N(\epsilon)$, then

$$\left|P_{ij}^n - A_{ij}\right| < \epsilon \quad \text{for all } i, j.$$

We give the intuition behind the next corollary. Suppose we begin in state i and we want to know the probability we are in state j if we let the process run for a long period of time. That is, we want to know

$$\lim_{n\to\infty} P_{ij}^n.$$

If it is possible to go from any state to any state—and this is the case if P is positive—then it should not matter in the long run where we begin. That is,

$$\lim_{n\to\infty} P_{ij}^n$$

should be the same for all i. So if

$$\lim_{n\to\infty} P^n = A,$$

then for j fixed, A_{ij} are the same for all i; that is, for a given column of A, the entries in that column are identical. This means that the rows of A are the same. So we have the following corollaries:

Corollary

Let $\hat{a} = (a_1, \ldots, a_n)$ be one of the identical rows of A. If $\hat{p} = (p_1, \ldots, p_n)$ is any probability vector, then

$$\lim_{n\to\infty} \hat{p} P^n = \hat{a}.$$

Proof

We have

$$\begin{aligned}
\hat{p}A &= (p_1, \ldots, p_n)\begin{pmatrix} a_1 & \cdots & a_n \\ \vdots & & \vdots \\ a_1 & \cdots & a_n \end{pmatrix} \\
&= (p_1a_1 + p_2a_1 + \cdots + p_na_1, p_1a_2 + p_2a_2 + \cdots + p_na_2, \ldots, p_1a_n + p_2a_n + \cdots + p_na_n) \\
&= ((p_1 + \cdots + p_n)a_1, (p_1 + \cdots + p_n)a_2, \ldots, (p_1 + \cdots + p_n)a_n) \\
&= (1\cdot a_1, 1\cdot a_2, \ldots, 1\cdot a_n) = (a_1, \ldots, a_n) = \hat{a}
\end{aligned}$$

and so

$$\lim_{n\to\infty} \hat{p} P^n = \hat{p} A = \hat{a}.$$

Example

For the transition matrix

$$P = \begin{pmatrix} .9 & .1 \\ .3 & .7 \end{pmatrix},$$

we have

$$P^{50} = \begin{pmatrix} .9 & .1 \\ .3 & .7 \end{pmatrix}^{50} \approx \begin{pmatrix} .75 & .25 \\ .75 & .25 \end{pmatrix} = A$$

and the unique equilibrium vector is (.75,.25). One can check that

$$(.75,.25)\begin{pmatrix} .9 & .1 \\ .3 & .7 \end{pmatrix} = (.75,.25).$$

Corollary

The vector $\hat{a}$ is the unique probability vector fixed by P.

Proof

Suppose $\hat{b}$ is a probability vector fixed by P; that is, $\hat{b}P = \hat{b}$. Then, for any positive integer n, we have $\hat{b}P^n = \hat{b}$, so

$$\hat{b} = \lim_{n\to\infty} \hat{b}P^n = \hat{b}A = \hat{a}.$$

The last theorem and its corollaries are very important in that they answer the fundamental question about the long-range behavior of finite Markov chains whose transition matrix is positive.

Recapping, for a Markov chain with positive transition matrix P

1. $\lim_{n\to\infty} P^n$ exists
2. If we denote the limit by matrix A, then
   (a) For any column of A, every entry is the same
   (b) Part (a) means that the rows of A are identical
   (c) Part (b) means that any initial probability evolves to the same probability distribution, which is the row of A

Said a slightly different way, the aforementioned results say that as far as finite-state Markov chains are concerned, if the transition matrix is positive, then we have an ideal situation. Namely, that any initial probability distribution of states evolves to the same probability distribution of states. We next show that the class of matrices for which this is true can be expanded.

Definition

A stochastic matrix P is called regular if there is a positive integer N for which P^N is positive.

If the transition matrix of a Markov process is regular and if P^N is positive, then it is possible to go between any two states in exactly N steps. This is more restrictive than saying it is possible to go between any two states as the matrix

$$P = \begin{pmatrix} 0 & 1 \\ 1 & 0 \end{pmatrix}$$

shows. In this case, it is possible for state i to go to the different state j in N steps if and only if N is odd, and state i to go to state i in N steps if and only if N is even.

Theorem 2.8

If P is a stochastic matrix and P^N is positive then P^{N+k} is positive for all $k \geq 0$.

Proof

Let P be a stochastic matrix and suppose P^N is positive. We show that P^{N+1} is positive. From this, it follows that P^{N+k} is positive for all $k \geq 0$.

Suppose

$$P = \begin{pmatrix} p_{11} & \cdots & p_{1n} \\ \vdots & & \vdots \\ p_{n1} & \cdots & p_{nn} \end{pmatrix} \quad \text{where } p_{ij} \geq 0 \text{ and } \sum_{j=1}^{n} p_{ij} = 1.$$

Let

$$P^N = \begin{pmatrix} b_{11} & \cdots & b_{1n} \\ \vdots & & \vdots \\ b_{n1} & \cdots & b_{nn} \end{pmatrix} \quad \text{where } b_{ij} > 0.$$

Then

$$\left(P^{N+1}\right)_{ij} = \sum_{k=1}^{n} \left(P^N\right)_{ik} P_{kj} = b_{i1}p_{1j} + \cdots + b_{in}p_{nj},$$

and since

$$p_{ij} \geq 0 \quad \text{and} \quad \sum_{j=1}^{n} p_{ij} = 1,$$

at least one $p_{ij} > 0$. In addition $b_{ij} > 0$, so

$$\left(P^{N+1}\right)_{ij} > 0.$$

It follows that P^{N+k} is positive for all $k \geq 0$.

We now consider the case where the matrix P is a regular stochastic matrix. We can mimic the proof of Theorem 2.6 to show that $M_{n+1} - m_{n+1} \leq M_n - m_n$ so that $\{M_n - m_n\}$ is a nonincreasing sequence. Since P^N satisfies the hypotheses of Theorem 2.6, we can conclude that the subsequence

$$\{M_{Nn} - m_{Nn}\}$$

converges to 0. From this, it follows that $\{M_n - m_n\}$ converges to 0, and the proof of Theorem 2.6 now applies to regular transition matrices.

Finding the Equilibrium State

If P be the transition matrix for a finite discrete Markov chain and P is a positive matrix or a regular matrix (thus ensuring there is a unique equilibrium state), then we can find the equilibrium state in two ways.

One method is to solve the matrix equation

$$\hat{x}P = \hat{x},$$

where $\hat{x}$ is a probability vector.

This matrix equation is easily converted into a system of linear equations, which any CAS program such as MAPLE® or MATLAB® will solve efficiently.

The other is by finding

$$\lim_{n\to\infty} P^n,$$

and the equilibrium state will be the common row of the limiting matrix.

Example

We find the equilibrium state and the limiting matrix for

$$P=\begin{pmatrix} \frac{1}{4} & 0 & \frac{1}{8} & \frac{5}{8} \\ \frac{1}{3} & \frac{1}{3} & \frac{1}{6} & \frac{1}{6} \\ \frac{1}{10} & \frac{1}{5} & \frac{1}{5} & \frac{1}{2} \\ \frac{1}{9} & \frac{2}{9} & \frac{4}{9} & \frac{2}{9} \end{pmatrix}.$$

Let $\hat{x}=(a,b,c,d)$. We solve

$$\hat{x}=\hat{x}P;\text{that is } (a,b,c,d)=(a,b,c,d)\begin{pmatrix} \frac{1}{4} & 0 & \frac{1}{8} & \frac{5}{8} \\ \frac{1}{3} & \frac{1}{3} & \frac{1}{6} & \frac{1}{6} \\ \frac{1}{10} & \frac{1}{5} & \frac{1}{5} & \frac{1}{2} \\ \frac{1}{9} & \frac{2}{9} & \frac{4}{9} & \frac{2}{9} \end{pmatrix}.$$

Now,

$$(a,b,c,d)\begin{pmatrix} \frac{1}{4} & 0 & \frac{1}{8} & \frac{5}{8} \\ \frac{1}{3} & \frac{1}{3} & \frac{1}{6} & \frac{1}{6} \\ \frac{1}{10} & \frac{1}{5} & \frac{1}{5} & \frac{1}{2} \\ \frac{1}{9} & \frac{2}{9} & \frac{4}{9} & \frac{2}{9} \end{pmatrix}$$

$$=\left(\frac{a}{4}+\frac{b}{3}+\frac{c}{10}+\frac{d}{9},\frac{b}{3}+\frac{c}{5}+\frac{2d}{9},\frac{a}{8}+\frac{b}{6}+\frac{c}{5}+\frac{4d}{9},\frac{5a}{8}+\frac{b}{6}+\frac{c}{2}+\frac{2d}{9}\right).$$

So. we have the system of equations

$$a=\frac{a}{4}+\frac{b}{3}+\frac{c}{10}+\frac{d}{9}$$

$$b=\frac{b}{3}+\frac{c}{5}+\frac{2d}{9}$$

$$c = \frac{a}{8} + \frac{b}{6} + \frac{c}{5} + \frac{4d}{9}$$

$$d = \frac{5a}{8} + \frac{b}{6} + \frac{c}{2} + \frac{2d}{9},$$

to which we must add

$$a + b + c + d = 1,$$

because $\hat{x} = (a,b,c,d)$ is a probability vector.

Using a CAS, the solution to the system of five equations is

$$a = \frac{192}{1085} \approx .1769585253, \quad b = \frac{216}{1085} \approx .1990783410,$$

$$c = \frac{58}{217} \approx .2672811059, \quad d = \frac{387}{1085} \approx .3566820276.$$

Checking that this gives the same answer as

$$\lim_{n\to\infty} P^n,$$

we find that the CAS gives

$$P^{100} = \begin{pmatrix} .1769585253 & .1990783410 & .267281105 & .3566820276 \\ .1769585253 & .1990783410 & .267281105 & .3566820276 \\ .1769585253 & .1990783410 & .267281105 & .3566820276 \\ .1769585253 & .1990783410 & .267281105 & .3566820276 \end{pmatrix}.$$

Transient and Recurrent States

Definition

State i is recurrent if beginning in state i, the process returns to state i at some later time with probability 1, and state i is transient if beginning in state i, there is a positive probability the process will never return to state i.

Said another way, in a Markov chain, state i is said to be recurrent if

$$P\left(X_n = \text{i for some } n \geq 1 | X_0 = i\right) = 1$$

and transient if

$$P(X_n = \text{i for some } n \geq 1 | X_0 = i) < 1.$$

If the process is ever in the state i and i is recurrent, then the process returns to state i with probability 1. Once it returns (by the Markov property and time homogeneity), it is *starting over* and will again return to state i with probability 1. This happens infinitely often, so the expected number of returns is infinite.

We now compute the expected number of times that the process visits a transient state.

Let

f_{ij} = the probability of visiting the state j having begun in the state i

f_{jj} = the probability of returning to the state j having begun in the state j.

Thus, the state j is recurrent if $f_{jj} = 1$ and transient if $f_{jj} < 1$.

Also let

$$N_j = \text{the number of visits to state } j.$$

Theorem 2.9

Suppose state j is transient. Then

(a) $P\{N_j = m | X_0 = j\} = f_{jj}^{m-1}(1 - f_{jj})$.

(b) If $i \neq j$, then $P\{N_j = m | X_0 = i\} = \begin{cases} 1 - f_{ij} & \text{if } m = 0 \\ f_{ij} f_{jj}^{m-1}(1 - f_{jj}) & \text{if } m \geq 1. \end{cases}$

Proof

(a) In this case $m \geq 1$. If $X_0 = j$, to get m visits to state j there must be $m - 1$ return visits, which occurs with probability f_{jj}^{m-1} and after the $(m - 1)$st return, there must not be an additional return, which occurs with probability $(1 - f_{jj})$.

(b) In this case, it is possible that there are no visits to state j and this occurs with probability $(1 - f_{ij})$. If the process does visit state j, the first visit occurs with probability f_{ij}. If there are to be $m \geq 1$ visits, then after the first visit, there must be $m - 1$ return visits, which occurs with probability f_{jj}^{m-1}. After the mth visit, there must be no additional returns, which occurs with probability $(1 - f_{jj})$.

Corollary

If j is a transient state, then

$$E\{N_j | X_0 = j\} = \sum_{m=1}^{\infty} m f_{jj}^{m-1}(1 - f_{jj}) = (1 - f_{jj}) \sum_{m=1}^{\infty} m f_{jj}^{m-1}$$

$$= (1 - f_{jj}) \sum_{m=1}^{\infty} m f_{jj}^{m-1} = \frac{1}{(1 - f_{jj})}.$$

Proof

We have

$$\sum_{k=1}^{N} \frac{d}{dx} x^k = \sum_{k=1}^{N} k x^{k-1}$$

and if $|x| < 1$, then

$$\sum_{k=1}^{\infty} \frac{d}{dx} x^k = \frac{d}{dx}\left(\sum_{k=1}^{\infty} x^k\right) = \frac{d}{dx}\left(\frac{x}{1-x}\right) = \frac{1}{(1-x)^2}.$$

Thus,

$$(1-x) \sum_{k=1}^{\infty} k x^{k-1} = (1-x) \cdot \frac{1}{(1-x)^2} = \frac{1}{(1-x)}.$$

The result follows by taking $x = f_{jj}$.

Corollary

If j is a transient state, then

$$\text{If } i \neq j, \quad \text{then } E\{N_j | X_0 = i\} = \frac{f_{ij}}{(1 - f_{jj})}.$$

Proof

$$\text{If } i \neq j, \quad \text{then } E\{N_j | X_0 = i\} = \sum_{m=1}^{\infty} m f_{ij} f_{jj}^{m-1}(1 - f_{jj})$$

$$= f_{ij}(1 - f_{jj}) \sum_{m=1}^{\infty} m f_{jj}^{m-1} = \frac{f_{ij}}{(1 - f_{jj})}.$$

Indicator Functions

Indicator random variables can be a useful tool to compute expected values when the range of the function is a subset of the positive integers as we demonstrate in the proof of the next theorem.

An indicator random variable is a function of the form

$$I\{A\} \text{ or } I_{\{A\}} = \begin{cases} 1 & \text{if } A \text{ occurs} \\ 0 & \text{if } A \text{ does not occur} \end{cases}.$$

This means

$$E\left[I_{\{A\}}\right] = 0 \cdot P\left(A^c\right) + 1 \cdot P(A) = P(A),$$

which is an often-used fact. A simple way to determine whether a state is transient or recurrent is given by the next theorem.

Theorem 2.10

Starting in state i, the expected number of returns to state i is

$$\sum_{n=0}^{\infty} p_n(i,i).$$

Thus, state i is transient if and only if

$$\sum_{n=0}^{\infty} p_n(i,i) < \infty$$

and so is recurrent if and only if

$$\sum_{n=0}^{\infty} p_n(i,i) = \infty.$$

Proof

Let

$$I_n = \begin{cases} 1 & \text{if } X_n = i, \text{ given } X_0 = i \\ 0 & \text{if } X_n \neq i, \text{ given } X_0 = i \end{cases}$$

and

$$N = \sum_{n=1}^{\infty} I_n.$$

Then N is the number of times that the process returns to the state i having begun in state i, so state i is transient if and only if

$$E\left[N \middle| X_0 = i\right] < \infty.$$

Now

$$E\left[N \middle| X_0 = i\right] = E\left[\sum_{n=1}^{\infty} I_n \middle| X_0 = i\right] = \sum_{n=1}^{\infty} E\left[I_n \middle| X_0 = i\right].$$

But

$$E\left[I_n \middle| X_0 = i\right] = P\left\{I_n = 1 \middle| X_0 = i\right\} = P\left\{X_n = i \middle| X_0 = i\right\} = p_n\left(i,i\right),$$

so

$$E\left[N \middle| X_0 = i\right] = \sum_{n=0}^{\infty} p_n\left(i,i\right)$$

and the result follows.

Corollary

If j is a transient state, then

$$\lim_{n \to \infty} p_n\left(i,j\right) = 0$$

for all states i.

Proof
Let

$$r_{ij} = E\{N_j | X_0 = i\}.$$

If j is a transient state and $i \neq j$, then

$$r_{ij} = E\{N_j | X_0 = i\} = \frac{f_{ij}}{(1 - f_{jj})}$$

and if $i = j$ then

$$r_{jj} = E\{N_j | X_0 = j\} = \frac{1}{(1 - f_{jj})}$$

so that

$$r_{ij} = r_{jj} f_{ij}.$$

We also have

$$r_{ij} = E\{N_j | X_0 = i\} = \sum_{n=1}^{\infty} p_n(i, j),$$

so if r_{jj} is finite, then r_{ij} is finite and

$$\sum_{n=1}^{\infty} p_n(i, j)$$

converges. Thus,

$$\lim_{n \to \infty} p_n(i, j) = 0.$$

Corollary

In a finite-state Markov chain, not all states are transient.

Proof

Suppose that the states in a Markov chain are 1, 2, …, N and all are transient. Then for each state i, there is a time T_i after which there is a positive probability state i will never be revisited. If $T = T_1 + \cdots + T_N$, then after time T, there would be a positive probability the process is not in any state.

Theorem 2.11

If state i is recurrent and j is a state for which there is a positive integer q with $p_q(i, j) > 0$, then

(a) there is a positive integer r for which $p_r(j, i) > 0$
(b) the state j is recurrent

Proof

(a) We prove the contrapositive of the theorem.
If there is not a positive integer r for which $p_r(j, i) > 0$, then once the process is in state j, it will never return to state i. Thus, there is a positive probability the process would never return to state i and state i would be transient.
(b) We show state j is recurrent by showing

$$\sum_{m=1}^{\infty} p_m(j,j) = \infty.$$

For n a positive integer, we have

$$p_{q+n+r}(j,j) = \sum_{k \in S} p_q(j,k)\,p_n(k,k)\,p_r(k,j) \geq p_q(j,i)\,p_n(i,i)\,p_r(i,j),$$

so

$$\sum_{n=1}^{\infty} p_{q+n+r}(j,j) \geq \sum_{n=1}^{\infty} p_q(j,i)\,p_n(i,i)\,p_r(i,j) = p_q(j,i)\,p_r(i,j)\sum_{n=1}^{\infty} p_n(i,i) = \infty.$$

Theorem 2.12

In a Markov chain with N transient states, there is a positive probability that the process will move from a transient state to a recurrent state in N or fewer steps.

Proof

Label the transient states 1, 2, …, N. Suppose that the process begins in one of these states, call it i. It must be the case that there is a positive probability to go from some transient states to a recurrent state in one step. Beginning in state i, the process must eventually visit one of these states or one of the transient states would have to be visited infinitely often, contrary to being transient.

It must be possible to go from state i to this step in no more than $N - 1$ steps, because otherwise, some state would have been visited twice, and the path would have a loop that could be eliminated.

Later in this chapter, we will study absorbing Markov chains. A state in a Markov chain is absorbing if once the process enters that state, it never leaves it. A Markov chain is absorbing if beginning at any state, there is a positive probability the process will eventually enter an absorbing state. An example of such a Markov chain is the gambler's ruin problem. The next corollary will be used in analyzing absorbing Markov chains.

Corollary

(a) Let $\{X_n\}$ be a Markov chain with N transient states. Recall that a finite Markov chain cannot consist entirely of transient states. Enumerate the states so that 1, …, N are the transient states and let Q be the matrix consisting of the first N rows and first N columns of P, where P is the transition matrix of the Markov chain. Then Q^N is a nonnegative matrix for which, the sum of each row is less than 1.

(b) Let r_i be the sum of the entries of the ith row of Q^N and let $r = \max\{r_1, \ldots, r_N\}$. Then $r < 1$ and the sum of any row of $Q^{kN} \leq r^k$.

(c) The entries of $Q^n \to 0$ uniformly as $n \to \infty$; that is, given $\epsilon > 0$, there is an $N(\epsilon)$ so that if $n > N(\epsilon)$, then $Q^n(i, j) < \epsilon$ for $1 \leq i, j \leq n$.

Proof

(a) Since it is possible for every transition state to evolve to a recurrent state after N steps, the sum of every row of Q^N is less than 1.

(b) We give the proof of part (b) in the case $k = 2$ and leave the proof of the remainder as an exercise.

To simplify the notation, let $A = Q^N$. Consider the entries in the lth row of A^2. We have

$$\left(A^2\right)_{lj} = \sum_{i=1}^{N} A_{li} A_{ij} \quad \text{and} \quad \text{so} \sum_{j=1}^{N} \left(A^2\right)_{lj} = \sum_{j=1}^{N} \left(\sum_{l=1}^{N} A_{li} A_{ij} \right)$$

is the sum of the entries of the lth row of A^2, and we want to show

$$\sum_{j=1}^{N}\left(\sum_{l=1}^{N} A_{li}A_{ij}\right) \le r^2.$$

Now

$$\sum_{j=1}^{N}\left(\sum_{l=1}^{N} A_{li}A_{ij}\right) = \sum_{l=1}^{N}\left(\sum_{j=1}^{N} A_{li}A_{ij}\right) = \sum_{j=1}^{N} A_{ij}\left(\sum_{l=1}^{N} A_{li}\right)$$

and

$$\sum_{l=1}^{N} A_{li} \le r,$$

so

$$\sum_{j=1}^{N} A_{ij}\left(\sum_{l=1}^{N} A_{li}\right) \le \sum_{j=1}^{N} A_{ij} r = r\sum_{j=1}^{N} A_{ij} \le r^2$$

since

$$\sum_{j=1}^{N} A_{ij} \le r.$$

(c) Part (c) follows immediately from part (b)

We note that an alternate proof for part (b) is that by Theorem 2.6, the maximum eigenvalue of Q^N is less than 1. If ρ is this maximum eigenvalue, then the maximum eigenvalue of Q^{Nk} is ρ^k and

$$\lim_{k\to\infty} \rho^k = 0.$$

Perron–Frobenius Theorem

The Perron–Frobenius Theorem has many applications, among them predicting the long-range behavior of some Markov chains. Many of its uses come from the fact that for certain matrices, there is an eigenvector that will dominate a process which requires repeated application of the matrix.

Theorem 2.13 (Perron–Frobenius Theorem)

Suppose that is A an $n \times n$ positive matrix. Then A has an eigenvalue λ_{PF} for which

(i) λ_{PF} is real and positive
(ii) If λ is any other eigenvalue of A, then $|\lambda| < \lambda_{PF}$
(iii) The eigenvalue for λ_{PF} has algebraic (and thus geometric) dimension 1
(iv) λ_{PF} has a positive eigenvector

We do not give the proof of the theorem. The steps involved in a proof are given in Lawler (2006, pp. 40–41).

Recall that we have previously shown that 1 is an eigenvalue for a stochastic matrix, and it is the largest eigenvalue. The Perron–Frobenius Theorem ensures for a positive stochastic matrix the associated eigenspace of the eigenvalue 1 has dimension 1.

As an application of the Perron–Frobenius Theorem, we consider the following example.

Example

Suppose that

$$A = \begin{pmatrix} a_{11} & \cdots & a_{1n} \\ \vdots & & \vdots \\ a_{n1} & \cdots & a_{nn} \end{pmatrix}$$

is a positive transition matrix, and we want to find the stationary distribution for A. That is, we want to find the probability vector $\hat{x} = (x_1, \ldots, x_n)$, for which

$$\hat{x}A = \hat{x} \quad \text{or} \quad (x_1, \ldots, x_n)\begin{pmatrix} a_{11} & \cdots & a_{1n} \\ \vdots & & \vdots \\ a_{n1} & \cdots & a_{nn} \end{pmatrix} = (x_1, \ldots, x_n).$$

Taking the transpose of this equation, we get

$$\begin{pmatrix} a_{11} & \cdots & a_{n1} \\ \vdots & & \vdots \\ a_{1n} & \cdots & a_{nn} \end{pmatrix}\begin{pmatrix} x_1 \\ \vdots \\ x_n \end{pmatrix} = \begin{pmatrix} x_1 \\ \vdots \\ x_n \end{pmatrix}. \tag{1}$$

In other words, a stationary distribution is the transpose of the normalized eigenvector of A^T whose eigenvalue is 1. According to the Perron–Frobenius Theorem, if A is positive, this distribution exists and is unique. We describe how to find this distribution.

To solve Equation 1, we find a nonzero $\hat{x}$, for which

$$\left(A^T - I_n\right)\hat{x} = \hat{0}.$$

If we row reduce the matrix $A^T - I_n$, we get a matrix whose first $n - 1$ rows are not identically zero, and whose last row is all zeros. If we replace the last row with all 1s (which is the normalizing equation for the vector $\hat{x}$), this yields the unique stationary distribution. The next example demonstrates the idea.

Example

For $0 < a, b < 1$, let

$$A = \begin{pmatrix} a & 1-a \\ b & 1-b \end{pmatrix}.$$

The restriction $0 < a, b < 1$ ensures that the matrix is positive. Now

$$A^T - I_2 = \begin{pmatrix} a & b \\ 1-a & 1-b \end{pmatrix} - \begin{pmatrix} 1 & 0 \\ 0 & 1 \end{pmatrix} = \begin{pmatrix} a-1 & b \\ 1-a & -b \end{pmatrix}.$$

Then the matrix form of the equation

$$\left(A^T - I_n\right)\hat{x} = \hat{0}$$

is

$$\begin{pmatrix} a-1 & b \\ 1-a & -b \end{pmatrix}\begin{pmatrix} x_1 \\ x_2 \end{pmatrix} = \begin{pmatrix} 0 \\ 0 \end{pmatrix}.$$

To solve this, we row reduce

$$\begin{pmatrix} a-1 & b & 0 \\ 1-a & -b & 0 \end{pmatrix}$$

and get

$$\begin{pmatrix} 1 & \dfrac{b}{a-1} & 0 \\ 0 & 0 & 0 \end{pmatrix}.$$

We want the vector to be a probability vector, so next, we replace the row of 0s by a row of 1s (since we need the equation $x + y = 1$) to get

$$\begin{pmatrix} 1 & \dfrac{b}{a-1} & 0 \\ 1 & 1 & 1 \end{pmatrix}.$$

When this matrix is row reduced, the result is

$$\begin{pmatrix} 1 & 0 & \dfrac{-b}{a-b-1} \\ 0 & 1 & \dfrac{a-1}{a-b-1} \end{pmatrix},$$

so

$$(x_1, x_2) = \left(\frac{-b}{a-b-1}, \frac{a-1}{a-b-1} \right)$$

is the unique probability vector for which $\hat{x}A = \hat{x}$.

To give a specific case, suppose that

$$A = \begin{pmatrix} .9 & .1 \\ .3 & .7 \end{pmatrix}$$

so that $a = .9$ and $b = .3$. The example asserts that the equilibrium distribution is

$$\left(\frac{-.3}{.9-.3-1}, \frac{.9-1}{.9-.3-1} \right) = \left(\frac{3}{4}, \frac{1}{4} \right)$$

and, in fact, we note

$$\left(\frac{3}{4}, \frac{1}{4} \right) \begin{pmatrix} .9 & .1 \\ .3 & .7 \end{pmatrix} = \left(\frac{3}{4}, \frac{1}{4} \right).$$

Theorem 2.14

If in addition to being positive, the transition matrix has a basis of eigenvectors, the rate of convergence to the equilibrium state is exponential.

Proof

Suppose that $\{\hat{x}_1, \ldots, \hat{x}_n\}$ is a basis of eigenvectors for P with $P\hat{x}_i = \lambda_i \hat{x}_i$, $i = 1, \ldots, n$ with $\lambda_1 = 1$ and $|\lambda_i| < 1, i = 2, \ldots, n$. Also, suppose that

$$\hat{x} = a_1\hat{x}_1 + a_2\hat{x}_2 + \cdots + a_n\hat{x}_n.$$

Then

$$P^k\hat{x} = P^k\left(a_1\hat{x}_1 + a_2\hat{x}_2 + \cdots + a_n\hat{x}_n\right) = a_1{\lambda_1}^k\hat{x}_1 + a_2{\lambda_2}^k\hat{x}_2 + \cdots + a_n{\lambda_n}^k\hat{x}_n$$

$$= a_1\hat{x}_1 + a_2{\lambda_2}^k\hat{x}_2 + \cdots + a_n{\lambda_n}^k\hat{x}_n.$$

Since $|\lambda_i| < 1$, $i = 2, \ldots, n$, we have ${\lambda_2}^k\hat{x}_2 + \cdots + a_n{\lambda_n}^k\hat{x}_n$, which converges to $\hat{0}$ exponentially, so

$$\lim_{k\to\infty} P^k\hat{x} = a_1\hat{x}_1.$$

In fact, we have

$$\left\|P^k\hat{x} - a_1\hat{x}_1\right\| = \left\|(a_1\hat{x}_1 + a_2{\lambda_2}^k\hat{x}_2 + \cdots + a_n{\lambda_n}^k\hat{x}_n) - a_1\hat{x}_1\right\| = \left\|a_2{\lambda_2}^k\hat{x}_2 + \cdots + a_n{\lambda_n}^k\hat{x}_n\right\|,$$

where $\|\hat{v}\|$ is the length of the vector $\hat{v}$.

Let λ_j be the eigenvalue whose modulus is the largest of the moduli of $\lambda_2, \ldots, \lambda_n$. Then $|\lambda_j| < 1$, and

$$\left\|a_2{\lambda_2}^k\hat{x}_2 + \cdots + a_n{\lambda_n}^k\hat{x}_n\right\| \le |\lambda_2|^k \left\|a_2\hat{x}_2\right\| + \cdots + |\lambda_n|^k \left\|a_n\hat{x}_n\right\| \le |\lambda_j|^k \left(\left\|a_2\hat{x}_2\right\| + \cdots + \left\|a_n\hat{x}_n\right\|\right).$$

Let $K = \|a_2\hat{x}_2\| + \cdots + \|a_n\hat{x}_n\|$. Then $\left\|P^k\hat{x} - a_1\hat{x}_1\right\| \le K|\lambda_j|^k$, so we have shown that the equilibrium state is $a_1\hat{x}_1$, and the rate of convergence is exponential.

The convergence to $\hat{0}$ noted earlier is true even if there is not a basis of eigenvectors, but the reasons are deeper. Before explaining these reasons, we give a heuristic explanation of what occurs.

If 1 is an eigenvalue of the linear transformation A with algebraic dimension 1, then there is a basis for the vector space for which the matrix of A is of the form

$$\begin{pmatrix} 1 & 0 & 0 & \cdots & 0 \\ 0 & B_1 & & & \\ 0 & & B_2 & & \\ \vdots & & & \ddots & \\ 0 & & & & B_k \end{pmatrix}$$

where the B_i are block matrices, but not necessarily diagonal. This is the Jordan canonical form of the transformation.

If 1 is the largest eigenvalue of the matrix A, and all other eigenvalues λ have the property that $|\lambda| < 1$, then the blocks of A can be chosen so that $(B_i)^n$ goes to the zero matrix as n gets large. If the matrix is in this form, then

$$\lim_{n\to\infty}\begin{pmatrix} 1 & 0 & 0 & \cdots & 0 \\ 0 & B_1 & & & \\ 0 & & B_2 & & \\ \vdots & & & \ddots & \\ 0 & & & & B_k \end{pmatrix}^n = \begin{pmatrix} 1 & 0 & 0 & \cdots & 0 \\ 0 & 0 & & & \\ 0 & & 0 & & \\ \vdots & & & \ddots & \\ 0 & & & & 0 \end{pmatrix}.$$

We give the major ideas in another way to prove the claim.

The spectral radius of a matrix A, denoted $\rho(A)$, is defined by

$$\rho(A) = \max\{|\lambda| \mid \lambda \text{ is an eigenvalue of } A\}.$$

Fact: We have $\lim_{k\to\infty} A^k = 0$ if and only if $\rho(A) < 1$.

Let A be a positive stochastic matrix and let $\hat{v}$ be the unique positive eigenvector of A that has eigenvalue 1 guaranteed by the Perron–Frobenius Theorem. Add vectors to $\hat{v}$ to complete a basis. Suppose the basis is $\{\hat{v}, \hat{x}_2, \ldots, \hat{x}_n\}$. By the Perron–Frobenius Theorem, any eigenvalue of A restricted to the vector space generated by $\{\hat{x}_2, \ldots, \hat{x}_n\}$ has modulus less than 1. Let A^* denote A restricted to the vector space generated by $\{\hat{x}_2, \ldots, \hat{x}_n\}$. Then

$$\rho(A^*) < 1.$$

Now let $\hat{w}$ be a probability vector in the original vector space. Then

$$\hat{w} = c_1\hat{v} + c_2\hat{x}_2 + \cdots + c_n\hat{x}_n$$

$$\begin{aligned} A^k(\hat{w}) &= A^k(c_1\hat{v} + c_2\hat{x}_2 + \cdots + c_n\hat{x}_n) = A^k(c_1\hat{v}) + A^k(c_2\hat{x}_2 + \cdots + c_n\hat{x}_n) \\ &= (c_1\hat{v}) + A^k(c_2\hat{x}_2 + \cdots + c_n\hat{x}_n) \to c_1\hat{v} \quad \text{as } k \to \infty. \end{aligned}$$

Absorbing Markov Chains

Definition

State i in a Markov chain is absorbing if once the process enters that state, it never leaves. This is true if and only if $P_{ii} = 1$. If it is possible to go from every state to an absorbing state (not necessarily in one step), the process is said to be an absorbing Markov chain.

Thus, in an absorbing Markov chain, every state is either transient or absorbing.

To study absorbing Markov chains, it will be convenient to arrange the states so that the transition matrix has a particular form. To that end, suppose that we have an absorbing Markov chain of n states and that there are k absorbing states (and thus $n - k$ non-absorbing states). Number the states so that 1, …, k are the absorbing states. Then the first k rows of the transition matrix are

$$
\begin{matrix}
1 & 0 & 0 & 0 & 0 & \cdots & 0 \\
0 & 1 & 0 & 0 & 0 & \cdots & 0. \\
\vdots & \vdots & \vdots & \vdots & \vdots & \vdots & \vdots \\
\\
0 & 0 & 0 & 1 & 0 & \cdots & 0
\end{matrix}
$$

That is, we have the $k \times k$ identity matrix beside the $k \times (n - k)$ zero matrix. We will find it convenient to write this as

$$\mathbf{I0},$$

where
I is the $k \times k$ identity matrix
0 is the $k \times (n - k)$ zero matrix

We express the remainder of the rows of the transition matrix as

$$\mathbf{RQ},$$

where
R is an $(n - k) \times k$ matrix
Q is an $(n - k) \times (n - k)$ matrix

If we number the rows and columns of **R** and **Q** as they will be in the transition matrix, then the range of values of i and j in $\mathbf{R}_{ij}$ is $k + 1 \le i \le n, 1 \le j \le k$ and the range of values of i and j in $\mathbf{Q}_{ij}$ is $k + 1 \le i, j \le n$.

With this convention, $\mathbf{R}_{ij}$ is the probability of going from the non-absorbing state i to the absorbing state j in one step and $\mathbf{Q}_{ij}$ is the probability of going from the non-absorbing state i to the non-absorbing state j in one step.

Example

Consider the gambler's ruin problem where there are two players, Arnie and Bill. We number the states according to Arnie's fortune, 0, 1, 2, 3, 4. Suppose that Arnie wins a single play of the game with probability .6. The transition matrix for the original matrix is given in Figure 2.4, and the transition matrix for the renumbered states is given in Figure 2.5.

The absorbing states are 0 and 4. We renumber the states as follows:

Old Numbering	New Numbering
0	0
1	2
2	3
3	4
4	1

In this case, constructing the transition matrix with the renumbered states from the transition matrix with the original numbering was simple enough to be done by inspection. If that is not the case, it can be done by an algorithm that we now demonstrate.

$$P = \begin{pmatrix} 1 & 0 & 0 & 0 & 0 \\ .4 & 0 & .6 & 0 & 0 \\ 0 & .4 & 0 & .6 & 0 \\ 0 & 0 & .4 & 0 & .6 \\ 0 & 0 & 0 & 0 & 1 \end{pmatrix}$$

FIGURE 2.4
Transition for the original Markov chain.

$$P^* = \begin{pmatrix} 1 & 0 & 0 & 0 & 0 \\ 0 & 1 & 0 & 0 & 0 \\ .4 & 0 & 0 & .6 & 0 \\ 0 & 0 & .4 & 0 & .6 \\ 0 & .6 & 0 & .4 & 0 \end{pmatrix}$$

FIGURE 2.5
Transition for the Markov chain with renumbered states.

In this example, consider the states to be numbered 1 through 5, with state 1 representing Arnie's fortune of \$0, state 2 his fortune of \$1, state 3 his fortune of \$2, state 4 his fortune of \$3, and state 5 his fortune of \$4.

In the renumbered system, state 1 is a fortune of \$0, state 2 is a fortune of \$4, state 3 a fortune of \$1, state 4 a fortune of \$2, and state 5 a fortune of \$3.

Let

$$f : \text{Old State} \to \text{New State}.$$

In this case, $f(1) = 1,\ f(2) = 3,\ f(3) = 4, f(4) = 3, f(5) = 2$.

To find the entries of P^*, the (i, j) entry of P becomes the $\left(f(i), f(j)\right)$ entry of P^*.

So, for example, the (3, 1) entry of P^* is the (2, 1) entry of P. (Check that both are .4.)

One could also say that the (i, j) entry of P^* is the $\left(f^{-1}(i), f^{-1}(j)\right)$ entry of P.

Returning to the example, the matrices **I**, **O**, **Q**, and **R** for the renumbered case are

$$\mathbf{I} = \begin{pmatrix} 1 & 0 \\ 0 & 1 \end{pmatrix},\quad \mathbf{O} = \begin{pmatrix} 0 & 0 & 0 \\ 0 & 0 & 0 \end{pmatrix},\quad \mathbf{Q} = \begin{pmatrix} 0 & .6 & 0 \\ .4 & 0 & .6 \\ 0 & .4 & 0 \end{pmatrix},\quad \mathbf{R} = \begin{pmatrix} .4 & 0 \\ 0 & 0 \\ 0 & .6 \end{pmatrix}.$$

Let $\boldsymbol{P}^*$ denote the transition matrix in the renumbered system. Then we will write

$$\boldsymbol{P}^* = \begin{pmatrix} \mathbf{I} & \mathbf{0} \\ \mathbf{R} & \mathbf{Q} \end{pmatrix}.$$

If n is a positive integer, then

$$\left(\boldsymbol{P}^*\right)^n = \begin{pmatrix} \boldsymbol{I} & \mathbf{0} \\ \blacksquare & \boldsymbol{Q}^n \end{pmatrix},$$

where $\blacksquare$ is an $(n - k) \times k$ matrix.

Note that since it is possible to go from every state to an absorbing state, the matrix $\mathbf{Q}^n$ goes to 0 as n goes to infinity, as we showed in a corollary to Theorem 2.13.

In the example earlier, the entries of $\mathbf{Q}^{20}$ are on the order of 10^{-8}.

The matrix **Q** should be thought of as giving the probability of going from one non-absorbing state to another. So $\mathbf{Q}_{ij}$ gives the probability of going from the non-absorbing state i to the non-absorbing state j in one step, and $(\mathbf{Q}^n)_{ij}$ gives the probability of going from the non-absorbing state i to the non-absorbing state j in exactly j steps.

The matrix **R** gives the probability of going from a non-absorbing state to an absorbing state. The number $\mathbf{R}_{ij}$ is the probability of going from the non-absorbing state i to the absorbing state j in one step.

Theorem 2.15

The probability that any state in a finite absorbing Markov chain is absorbed after n steps approaches 1 as n goes to infinity.

Proof

We show that the probability that any state in an absorbing Markov chain is not absorbed after n steps approaches 0 as n goes to infinity.

Let i be a non-absorbing state. Since the Markov state is absorbing, there is a positive integer m_i for which there is a positive probability p_i that i has moved to an absorbing state after m_i steps. Note that this implies that if $l \geq m_i$, then the probability that i has been absorbed after l steps is greater than or equal to p_i. Thus, the probability that state i has not been absorbed after m_i steps is less than or equal to $1 - p_i$.

Repeat the aforementioned procedure for each transient state, and let

$$m = \max\{m_i\}, \quad p = \min\{p_i\}.$$

Then, beginning in any state, the probability the process has not been absorbed after m steps is less than or equal to $1 - p$.

Thus, for any positive integer N, the probability the process has not been absorbed after Nm steps is less than or equal to $(1 - p)^N$ and

$$\lim_{N\to\infty}(1-p)^N = 0.$$

This also means that each entry of $\mathbf{Q}^n$ goes to 0 as n becomes large.

The next theorems deal with the series of matrices

$$\boldsymbol{I} + \boldsymbol{Q} + \boldsymbol{Q}^2 + \boldsymbol{Q}^3 + \cdots.$$

Before giving these results, we discuss why this is an important expression.

The probability that a process beginning in the non-absorbing state i is in the non-absorbing state j after n steps is $(Q^n)_{ij}$. Consider the gambler's ruin problem. Among the most important questions are

1. Given my beginning fortune, how likely is it that I will win or, said another way, how likely is it that my opponent will go bankrupt?
2. Given my beginning fortune, on average, how many games will we play before one of us goes bankrupt?

The answer to both questions depends on the series

$$\boldsymbol{I}+\boldsymbol{Q}+\boldsymbol{Q}^2+\boldsymbol{Q}^3+\cdots.$$

To develop some intuition for why this series is so important, we analyze the first question. The second question is likewise dependent on this series as we will see in Theorem 2.17.

Suppose that our initial state is i and we bankrupt our opponent when we reach state j. The ways this could happen are as follows:

(a) We win on the first play of the game. This means we went from the non-absorbing state i to the absorbing state j in one step. The probability of this occurring is $\boldsymbol{R}_{ij}$.

(b) We win on the second play of the game. So we went from the non-absorbing state i to some non-absorbing state k in the first play (which occurs with probability $\boldsymbol{Q}_{ik}$) and then went from the non-absorbing state k to the absorbing state j on the next play (which occurs with probability $\boldsymbol{R}_{kj}$). For a fixed non-absorbing state k, this sequence occurs with probability $\boldsymbol{Q}_{ik}\boldsymbol{R}_{kj}$. But this is true for any non-absorbing state k, so the probability that the game ends after exactly two plays is

$$\sum_{k \text{ nonabsorbing}} \boldsymbol{Q}_{ik}\boldsymbol{R}_{kj} = (\boldsymbol{QR})_{ij}.$$

(c) We win on the third play of the game. So we went from the non-absorbing state i to some non-absorbing state k in the first play (which occurs with probability $\boldsymbol{Q}_{ik}$) and then went from the non-absorbing state k to the absorbing state l on the next step (which occurs with probability $\boldsymbol{Q}_{kl}$), then went from the non-absorbing state l to the absorbing state j on the next play (which occurs with probability $\boldsymbol{R}_{lj}$). For particular non-absorbing states k and l, this occurs with probability $\boldsymbol{Q}_{ik}\boldsymbol{Q}_{kl}\boldsymbol{R}_{lj}$, but we must sum over all k and l to get

$$\sum_{k,l} \boldsymbol{Q}_{ik}\boldsymbol{Q}_{kl}\boldsymbol{R}_{lj} = (\boldsymbol{Q}^2\boldsymbol{R})_{ij}.$$

A pattern is emerging that if we are going to go from the non-absorbing state i to the absorbing state j for the first time $N + 1$ plays of the game, then we must have the first N plays where we stay in absorbing states and in the last play, we move to the absorbing state j and the probability of this occurring is $(\boldsymbol{Q}^N\boldsymbol{R})_{ij}$.

These events are mutually exclusive and exhaustive. Thus the probability we win is

$$\boldsymbol{R}_{ij} + (\boldsymbol{Q}\boldsymbol{R})_{ij} + (\boldsymbol{Q}^2\boldsymbol{R})_{ij} + (\boldsymbol{Q}^3\boldsymbol{R})_{ij} + \cdots.$$

Assuming that the necessary manipulations with infinite sums are valid (and we will show that they are), this can be expressed as

$$\left[\boldsymbol{R}\sum_{n=0}^{\infty}\boldsymbol{Q}^n\right]_{ij}.$$

NOTE: In subsequent discussions, when the identity matrix is used in combination with the matrix **Q**, we will denote the matrix as **I**. That is, whenever the identity matrix is used, it will be understood to be the proper dimension.

Theorem 2.16

For a finite absorbing Markov chain, let **Q** be the matrix above. Let **I** be the identity matrix with the same dimensions as **Q**. Then the matrix **I** – **Q** is invertible and

$$(\mathbf{I}-\mathbf{Q})^{-1} = \mathbf{I} + \mathbf{Q} + \mathbf{Q}^2 + \mathbf{Q}^3 + \cdots.$$

Note the analogy of this with geometric series of numbers. Namely, if $|x| < 1$ then

$$(1-x)^{-1} = \frac{1}{1-x} = \sum_{n=0}^{\infty} x^n = 1 + x + x^2 + \cdots.$$

Proof

The matrix $\mathbf{I}-\mathbf{Q}$ is invertible if and only if the only solution to

$$(\mathbf{I}-\mathbf{Q})\hat{x} = \hat{0} \quad \text{is} \quad \hat{x} = \hat{0}.$$

Now if $(\mathbf{I}-\mathbf{Q})\hat{x}=\hat{0}$, then $\mathbf{I}\hat{x}=\mathbf{Q}\hat{x}$, so $\hat{x}=\mathbf{Q}\hat{x}$ and it follows that for any positive integer n, $\hat{x}=\mathbf{Q}^n\hat{x}$. Thus,

$$\hat{x}=\lim_{n\to\infty}\mathbf{Q}^n\hat{x}=0\hat{x}=\hat{0},$$

so $\mathbf{I}-\mathbf{Q}$ is invertible.

To show

$$(\mathbf{I}-\mathbf{Q})^{-1}=\mathbf{I}+\mathbf{Q}+\mathbf{Q}^2+\mathbf{Q}^3+\cdots,$$

note that for any positive integer n, we have

$$(\mathbf{I}-\mathbf{Q})\left(\mathbf{I}+\mathbf{Q}+\mathbf{Q}^2+\mathbf{Q}^3+\cdots\mathbf{Q}^n\right)=\mathbf{I}-\mathbf{Q}^{n+1}$$

and

$$\lim_{n\to\infty}(\mathbf{I}-\mathbf{Q})\left(\mathbf{I}+\mathbf{Q}+\mathbf{Q}^2+\mathbf{Q}^3+\cdots\mathbf{Q}^n\right)=\lim_{n\to\infty}\mathbf{I}-\mathbf{Q}^{n+1}=\mathbf{I}.$$

Since

$$(\mathbf{I}-\mathbf{Q})\left(\mathbf{I}+\mathbf{Q}+\mathbf{Q}^2+\mathbf{Q}^3+\cdots\mathbf{Q}^n\right)=\left(\mathbf{I}+\mathbf{Q}+\mathbf{Q}^2+\mathbf{Q}^3+\cdots\mathbf{Q}^n\right)(\mathbf{I}-\mathbf{Q}),$$

it follows that

$$(\mathbf{I}-\mathbf{Q})^{-1}=\mathbf{I}+\mathbf{Q}+\mathbf{Q}^2+\mathbf{Q}^3+\cdots.$$

The matrix $(\mathbf{I}-\mathbf{Q})^{-1}$ is useful in computing several important characteristics for absorbing Markov chains. It is called the fundamental matrix and is denoted by $\boldsymbol{N}$.

Note that $\boldsymbol{N}$ can also be written as

$$N=\sum_{n=0}^{\infty}Q^n.$$

The next few results highlight some of the uses of $\boldsymbol{N}$.

Theorem 2.17

For a finite absorbing Markov chain with k non-absorbing states, the expected number of transitions that the non-absorbing state i undergoes before absorption is

$$\boldsymbol{N}_{i1} + \boldsymbol{N}_{i2} + \cdots + \boldsymbol{N}_{ik},$$

where

$$\boldsymbol{N} = (I - \boldsymbol{Q})^{-1} = I + \boldsymbol{Q} + \boldsymbol{Q}^2 + \boldsymbol{Q}^3 + \cdots.$$

This result can also be expressed in the following manner:

Let t_i be the expected time until absorption, beginning in the non-absorbing state i and let

$$\hat{t} = \begin{pmatrix} t_1 \\ \vdots \\ t_k \end{pmatrix}.$$

Then

$$\hat{t} = \boldsymbol{N} \begin{pmatrix} 1 \\ \vdots \\ 1 \end{pmatrix}.$$

Proof

Fix the non-absorbing state i and the non-absorbing state j. Let X_{in} be the random variable

$$X_{in} = \begin{cases} 1 & \text{if the system is in the state } j \text{ in } n \text{ steps having begun in state } i \\ 0 & \text{if the system is in any other state in } n \text{ steps having begun in } i \end{cases}.$$

Now

$$\Pr(X_{in} = 1) = \boldsymbol{Q}_{ij}^{n}; \quad \Pr(X_{in} = 0) = 1 - \boldsymbol{Q}_{ij}^{n},$$

so

$$E[X_{in}] = 0 \cdot \Pr(X_{in} = 0) + 1 \cdot \Pr(X_{in} = 1) = \boldsymbol{Q}_{ij}^{n}.$$

Then the expected number of visits to the transient state j that the state i undergoes before absorption is

$$\sum_{n=0}^{\infty} Q_{ij}^{n} = N_{ij}$$

and the expected total number of transitions that the process makes before absorption having begun in i is

$$\sum_{j=1}^{k} N_{ij} = N_{i1} + N_{i2} + \cdots + N_{ik},$$

which is the sum of the elements in the ith row of the matrix N.

Theorem 2.18

The probability that beginning in the non-absorbing state i the system is absorbed in the absorbing state j is given by

$$(NR)_{ij}.$$

NOTE: It is common in the literature to let the matrix NR be represented as B.

Proof

The proof is a repeat of the ideas of the gambler's ruin example.

We have

$$R_{kj} = \text{the probability of going from the non-absorbing state } k$$

to the absorbing state j in one step

$$Q_{ik}^{n} = \text{the probability of going from the non-absorbing state } i$$

to the non-absorbing state k in n steps, so

$$\sum_{k} Q_{ik}^{n} R_{kj}$$

(where the sum is over the non-absorbing states) is the probability of going from the non-absorbing state i to the absorbing state j in $n + 1$ steps and

$$\sum_{n=0}^{\infty}\left(\sum_{k} Q_{ik}^{n} R_{kj}\right)$$

is the probability that beginning in the non-absorbing state of i, the system is absorbed in the absorbing state j.

Now

$$\sum_{n=0}^{\infty}\left(\sum_{k} Q_{ik}^{n} R_{kj}\right)=\sum_{k}\left(\sum_{n=0}^{\infty} Q_{ik}^{n}\right) R_{kj}=\sum_{k} N_{ik} R_{kj}=(NR)_{ij},$$

so the probability that beginning in the non-absorbing state i the system is absorbed in the absorbing state j is given by

$$B_{ij}=(NR)_{ij}.$$

Example

We return to the gambler's ruin problem. Suppose that together the competitors have \$6. The states will be competitor A's fortune; that is, {0,1,…,6}. Suppose that A wins with probability .6. The transition matrix is then given by the probabilities

$$P(0,0)=1, P(6,6)=1, P(i,i+1)=.6, P(i,i-1)=.4; i=1,\ldots,5.$$

In the transition matrix, we arrange the states so that the absorbing states occupy the first two positions (with the state 0 in the first position) and states 1 through 5 occupy positions 3 through 7, respectively. Thus, the representation for P is

$$\begin{array}{l} \text{States} \\ \text{(cash)}\ 0\quad 6\quad 1\quad 2\quad 3\quad 4\quad 5 \end{array}$$

$$P=\begin{pmatrix} 1 & 0 & 0 & 0 & 0 & 0 & 0 \\ 0 & 1 & 0 & 0 & 0 & 0 & 0 \\ .4 & 0 & 0 & .6 & 0 & 0 & 0 \\ 0 & 0 & .4 & 0 & .6 & 0 & 0 \\ 0 & 0 & 0 & .4 & 0 & .6 & 0 \\ 0 & 0 & 0 & 0 & .4 & 0 & .6 \\ 0 & .6 & 0 & 0 & 0 & .4 & 0 \end{pmatrix}\begin{matrix} 0 \\ 6 \\ 1 \\ 2. \\ 3 \\ 4 \\ 5 \end{matrix}$$

The matrices $\boldsymbol{I}$, $\boldsymbol{R}$, and $\boldsymbol{Q}$ are

$$\boldsymbol{I}=\begin{pmatrix}1 & 0\\ 0 & 1\end{pmatrix} \quad \boldsymbol{R}=\begin{pmatrix}.4 & 0\\ 0 & 0\\ 0 & 0\\ 0 & 0\\ 0 & .6\end{pmatrix}$$

$$\boldsymbol{Q}=\begin{pmatrix}0 & .6 & 0 & 0 & 0\\ .4 & 0 & .6 & 0 & 0\\ 0 & .4 & 0 & .6 & 0\\ 0 & 0 & .4 & 0 & .6\\ 0 & 0 & 0 & .4 & 0\end{pmatrix}.$$

The matrix $\boldsymbol{N}$ is

$$\boldsymbol{N}=(\boldsymbol{I}-\boldsymbol{Q})^{-1}=\begin{pmatrix}211/133 & 195/133 & 9/7 & 135/133 & 81/133\\ 130/133 & 325/133 & 15/7 & 225/133 & 135/133\\ 4/7 & 10/7 & 19/7 & 15/7 & 9/7\\ 40/133 & 100/133 & 10/7 & 325/133 & 195/133\\ 16/133 & 40/133 & 4/7 & 130/133 & 211/133\end{pmatrix}$$

and

$$\begin{matrix} & \text{State}\,1(\$0) & \text{State}\,2(\$6)\end{matrix}$$
$$\boldsymbol{NR}=\begin{pmatrix}422/665 & 243/665\\ 52/133 & 81/133\\ 8/35 & 27/35\\ 16/133 & 117/133\\ 32/665 & 633/665\end{pmatrix}.$$

With these computations, we can answer some questions about the game.

1. Recall that N_{ij} is the expected number of times the process visits the transient state j having begun in the transient state i.
   With the renaming of the states, \$2 corresponds to the fourth state of the matrix P and \$4 corresponds to the sixth state of the matrix P. (This corresponds to the (2,4) entry of the matrices Q and N.) Thus, the expected number of times A's fortune is \$4 given that A started with \$2 is the (2, 4) entry of the matrix N which is 225/133.
2. Since $(\boldsymbol{NR})_{ij}$ is the probability that we are absorbed in the absorbing state j having begun in the transient state i, the probability that A wins having begun with \$2 is $(\boldsymbol{NR})_{42} = 81/133$. Here the (4,2) is the position in the 6×6 matrix.

3. The expected time before absorption is

$$\hat{t} = \mathbf{N}\hat{c},$$

where

$$\hat{c} = \begin{pmatrix} 1 \\ 1 \\ \vdots \\ 1 \end{pmatrix}$$

and t_i is the expected time before absorption having started in state i.

In this example

$$\hat{t} = \begin{pmatrix} \frac{211}{133} & \frac{195}{133} & \frac{9}{5} & \frac{135}{133} & \frac{81}{133} \\ \frac{130}{133} & \frac{325}{133} & \frac{15}{7} & \frac{225}{133} & \frac{135}{133} \\ \frac{4}{7} & \frac{10}{7} & \frac{19}{7} & \frac{15}{7} & \frac{9}{7} \\ \frac{40}{133} & \frac{100}{133} & \frac{10}{7} & \frac{325}{133} & \frac{195}{133} \\ \frac{16}{133} & \frac{40}{133} & \frac{4}{7} & \frac{130}{133} & \frac{211}{133} \end{pmatrix} \begin{pmatrix} 1 \\ 1 \\ 1 \\ 1 \\ 1 \end{pmatrix} = \begin{pmatrix} \frac{793}{133} \\ \frac{1100}{133} \\ \frac{57}{7} \\ \frac{850}{133} \\ \frac{473}{133} \end{pmatrix},$$

so the expected length of the game with A having started with \$3 is 57/7.

Thus, we have answered several of the most important questions about an absorbing Markov chain with relatively simple linear algebra.

Mean First Passage Time

We now move from absorbing Markov chains to a more general case.

Definition

An irreducible Markov chain is one where it is possible to go between any two states, not necessarily in one step.

Said another way, if P is the transition matrix for an irreducible Markov chain, then for any states i and j there is a positive integer n, which depends on i and j, for which $P^n(i,j) > 0$. While this is true if P is irreducible, this condition is less restrictive than saying there is a positive integer n for which P^n is a positive matrix. The matrix

$$P = \begin{pmatrix} 0 & 1 \\ 1 & 0 \end{pmatrix}$$

demonstrates this difference.

Definition

Suppose $\{X_n\}$ is an irreducible Markov chain with states i and j, $i \neq j$. The mean first passage time from i to j is the expected number of steps it takes to reach state j the first time having begun in state i.

One approach to finding the mean first passage time from i to j is to convert the original Markov chain to an absorbing Markov chain by converting state j to an absorbing state. This is done by changing the transition matrix P so that $P_{jj} = 1$ and $P_{jk} = 0$ if $k \neq j$.

If we start the process in any state i with $i \neq j$, then the process will evolve as in the original process until the state j is reached whereupon it is absorbed. The expected number of steps required can be found from the matrix **N** defined previously. We demonstrate this with an example.

Example

Let

$$P = \begin{pmatrix} 1/2 & 0 & 1/2 \\ 1/3 & 1/3 & 1/3 \\ 1/4 & 1/4 & 1/2 \end{pmatrix}$$

be the transition matrix for a Markov chain and suppose we want to find the expected number of steps until state 1 is reached the first time.

We change state 1 to an absorbing state and the transition matrix for the resulting chain is

$$P^* = \begin{pmatrix} 1 & 0 & 0 \\ 1/3 & 1/3 & 1/3 \\ 1/4 & 1/4 & 1/2 \end{pmatrix}.$$

So

$$Q^* = \begin{pmatrix} 1/3 & 1/3 \\ 1/4 & 1/2 \end{pmatrix}$$

and

$$N^* = \left(I - Q^*\right)^{-1} = \begin{pmatrix} 2 & 4/3 \\ 1 & 8/3 \end{pmatrix}.$$

To find the expected number of steps to go from a state other than 1 to state 1 for the first time, we compute

$$N^* \hat{c} = \begin{pmatrix} 2 & 4/3 \\ 1 & 8/3 \end{pmatrix} \begin{pmatrix} 1 \\ 1 \end{pmatrix} = \begin{pmatrix} 10/3 \\ 11/3 \end{pmatrix}.$$

This tells us that to go from state 2 to state 1, the expected number of steps is 10/3; to go from state 3 to state 1, the expected number of steps is 11/3.

Mean Recurrence Time and the Equilibrium State

Definition

The mean recurrence time for state i is the expected number of steps it takes to return to state i for the first time having begun in state i Thus, if

$$T_i = \min_{n \geq 1} \left\{ X_n = i \middle| X_0 = i \right\},$$

then $E[T_i]$ is the mean recurrence time for state i.

We want to relate the mean recurrence time of the states to the equilibrium state.

The main result of this section is that for an irreducible finite Markov chain if the mean recurrence time of state i is $E[T_i]$ and the equilibrium distribution is

$$\hat{\pi} = \left(\pi_1, \ldots, \pi_n\right), \text{then}$$

$$\frac{1}{E\left[T_i\right]} = \pi_i.$$

The proof of this fact is somewhat complicated, but the intuition is not. One way to view the equilibrium state is as a distribution that gives the long-run proportion of each state. Suppose we had a system and, on average, state 1 occurred every 10th step. Then it would seem that state 1 appeared 10% of the time. If state 1 occurred on every tenth step, then the mean occurrence time is 10. In this case, $E[T_1] = 10$ and also $\pi_1 = .10$, so

$$\frac{1}{E\left[T_1\right]} = \pi_1.$$

We already have two efficient ways of computing the equilibrium state of an irreducible Markov chain with transition matrix P, namely, finding

$$\lim_{n\to\infty} P^n$$

or finding the probability vector $\hat{\pi}$ for which

$$\hat{\pi}P = \hat{\pi}.$$

Thus, once we know

$$\frac{1}{E[T_i]} = \pi_i,$$

computing the mean first passage time is simple. The purpose of the analysis that follows is to validate what the intuition suggests about the relationship between the mean first passage time and the equilibrium state.

We elaborate on computing $E[T_i]$. Let

$$f_{ii}(n) = P\{X_n = i, X_{n-1} \neq i, \ldots, X_1 \neq i | X_0 = i\}, \quad n \geq 1,$$

so $f_{ii}(n)$ is the probability that if the process begins in state i, the first return to state i occurs on the nth step. Said another way,

$$f_{ii}(n) = P(T_i = n).$$

Thus,

$$E[T_i] = \sum_{n=1}^{\infty} n f_{ii}(n).$$

As a practical matter, $f_{ii}(n)$ can be difficult to determine. The following example demonstrates how to find $f_{ii}(n)$ and $E[T_i]$ for a two-state Markov chain.

Example

Suppose the transition matrix for a two-state Markov chain is

$$P = \begin{pmatrix} 1-a & a \\ b & 1-b \end{pmatrix},$$

where $0 < a, b < 1$. We designate the states as 1 and 2, and determine $E[T_1]$.

Now $f_{11}(1)$ is the probability that the process begins in the state 1 and remains in the state 1 on the first step, which is $P(1, 1) = 1 - a$.

Also $f_{11}(2)$ is the probability the process begins in the state 1, moves to state 2, and moves back to state 1 on the second step. So,

$$f_{11}(2) = P(1,2)P(2,1) = ab.$$

For $n \geq 3$, $f_{11}(n)$ can be illustrated with the diagram

$$1 \to 2 \to 2 \to \cdots \to 2 \to 1.$$

On the first step, the process moves from 1 to 2 (the probability of this is a). On the next $n - 2$ steps, the process remains in state 2; the probability of each of these steps is $1 - b$. On the nth step, the process moves from state 2 to state 1 with probability b. Thus, for $n \geq 3$,

$$f_{11}(n) = a(1-b)^{n-2}b,$$

so

$$E[T_1] = \sum_{n=1}^{\infty} nf_{11}(n) = 1 \cdot f_{11}(1) + 2 \cdot f_{11}(2) + \sum_{n=3}^{\infty} nf_{11}(n)$$

$$= 1(1-a) + 2ab + \sum_{n=3}^{\infty} na(1-b)^{n-2}b.$$

To compute the summand, we have

$$\sum_{n=3}^{\infty} na(1-b)^{n-2}b = ab(1-b)^{-1}\sum_{n=3}^{\infty} n(1-b)^{n-1}.$$

Now for $|x| < 1$, we have

$$\sum_{n=3}^{\infty} x^n = \frac{x^3}{1-x}$$

and

$$\frac{d}{dx}\left(\sum_{n=3}^{\infty} x^n\right) = \sum_{n=3}^{\infty} nx^{n-1}; \quad \frac{d}{dx}\left(\frac{x^3}{1-x}\right) = \frac{3x^2 - 2x^3}{(1-x)^2},$$

so

$$\sum_{n=3}^{\infty} nx^{n-1} = \frac{3x^2 - 2x^3}{(1-x)^2}.$$

Substituting $(1 - b)$ for x, we have

$$\sum_{n=3}^{\infty} n(1-b)^{n-1} = \frac{3(1-b)^2 - 2(1-b)^3}{\left[1-(1-b)\right]^2} = \frac{(1-b)^2(2b+1)}{b^2},$$

so

$$\sum_{n=3}^{\infty} na(1-b)^{n-2} b = ab(1-b)^{-1} \sum_{n=3}^{\infty} n(1-b)^{n-1}$$

$$= ab(1-b)^{-1} \frac{(1-b)^2(2b+1)}{b^2} = a\frac{(1-b)(2b+1)}{b}.$$

Thus,

$$E[T_1] = 1(1-a) + 2ab + \sum_{n=3}^{\infty} na(1-b)^{n-2} b = 1 - a + 2ab + a\frac{(1-b)(2b+1)}{b}. \quad (1)$$

We asserted earlier that this is $\frac{1}{\pi(1)}$ where $\pi = (\pi(1), \pi(2))$ is the equilibrium state. We take a specific example to corroborate this claim. Let

$$P = \begin{pmatrix} .7 & .3 \\ .6 & .4 \end{pmatrix},$$

so $a = .3$ and $b = .6$. We find the equilibrium state by raising P to a high power.

We have

$$P^{100} \approx \begin{pmatrix} \frac{2}{3} & \frac{1}{3} \\ \frac{2}{3} & \frac{1}{3} \end{pmatrix},$$

so

$$\pi(1) = \frac{2}{3}.$$

To find $E[T_1]$, we substitute $a = .3$ and $b = .6$ into expression (1). This gives

$$E[T_1] = 1 - (.3) + 2(.3)(.6) + (.3)\frac{(1-.6)(2 \cdot .6 + 1)}{.6} = 1.5,$$

so

$$\frac{1}{E[T_1]} = \frac{2}{3} = \pi(1).$$

In Exercise 2.16, we show

$$E[T_2] = \frac{1}{\pi(2)}.$$

We now formalize the connection between the mean recurrence time and the equilibrium state.

Theorem 2.18

For an irreducible finite state Markov chain, if the mean recurrence time of state i is $E[T_i]$ and the equilibrium distribution is

$$\hat{\pi} = (\pi_1, \ldots, \pi_n),$$

then

$$\frac{1}{E[T_i]} = \pi_i.$$

As we noted earlier, this is not an efficient way to compute the equilibrium state of an irreducible finite Markov chain. It does provide an efficient way to determine $E[T_i]$.

Proof

The proof of this is somewhat involved. Recall that the mean first passage time from i to j is the expected number of steps it takes to reach state j the first time having begun in the state i. To begin the proof, we define the mean first passage time matrix M by

$$M_{ij} = \begin{cases} \text{the mean first passage time from } i \text{ to } j \text{ if } i \neq j \\ 0 \text{ if } i = j \end{cases}.$$

We derive an equation for M_{ij}, where $i \neq j$.

The process moves from i to j in one step with probability P_{ij} (where P is the transition matrix).

If more than one step is required, the process moves from i to k, $k \neq j$ on the first step, and the expected number of steps to get from state k to state j is M_{kj}. So the total number of steps to move from i to j is $(M_{kj} + 1)$ and this occurs with probability P_{ik}. Thus,

$$M_{ij} = P_{ij} + \sum_{k \neq j} P_{ik} = \left(M_{kj} + 1\right) = P_{ij} + \sum_{k \neq j} P_{ik} M_{kj} + \sum_{k \neq j} P_{ik}$$

$$= \sum_{k} P_{ik} + \sum_{k \neq j} P_{ik} M_{kj} = 1 + \sum_{k \neq j} P_{ik} M_{kj}. \qquad (1)$$

Since $M_{jj} = 0$, we have

$$M_{ij} = 1 + \sum_{k \neq j} P_{ik} M_{kj} = 1 + \sum_{k} P_{ik} M_{kj} = 1 + (PM)_{ij} \text{ if } i \neq j. \qquad (2)$$

We now derive the analog to Equation 2 for the case $i = j$.

We denote the mean recurrence time of the state i by r_i. We derive an equation for r_i. The process returns to state i on the first step with probability P_{ii}.

If the process does not return to state i on the first step, then it moves to a state k, $k \neq i$ with probability P_{ik}. The process then moves from k to i, the expected number of steps from k to i being M_{ki}. There was the one additional step from k to i, so the total number of steps is $(M_{ki} + 1)$ and this occurs with probability P_{ik}. Thus,

$$r_i = P_{ii} + \sum_{k \neq i} P_{ik} \left(M_{ki} + 1\right) = \sum_{k} P_{ik} \left(M_{ki} + 1\right) = \left(\sum_{k} P_{ik} M_{ki}\right) + 1. \qquad (3)$$

Define the diagonal matrix D by

$$D_{ii} = r_i; \quad D_{ij} = 0 \text{ if } i \neq j.$$

Our goal now is to combine Equations 2 and 3 into a single matrix equation. To develop some intuition, we consider the case where there are three states. In this case

$$M = \begin{pmatrix} 0 & M_{12} & M_{13} \\ M_{21} & 0 & M_{23} \\ M_{31} & M_{32} & 0 \end{pmatrix} \quad D = \begin{pmatrix} r_1 & 0 & 0 \\ 0 & r_2 & 0 \\ 0 & 0 & r_3 \end{pmatrix}$$

$$PM = \begin{pmatrix} P_{11} & P_{12} & P_{13} \\ P_{21} & P_{22} & P_{23} \\ P_{31} & P_{32} & P_{33} \end{pmatrix} \begin{pmatrix} 0 & M_{12} & M_{13} \\ M_{21} & 0 & M_{23} \\ M_{31} & M_{32} & 0 \end{pmatrix}$$

$$= \begin{pmatrix} P_{12}M_{21} + P_{13}M_{31} & P_{11}M_{12} + P_{13}M_{32} & P_{11}M_{13} + P_{12}M_{23} \\ P_{22}M_{21} + P_{23}M_{31} & P_{21}M_{12} + P_{23}M_{32} & P_{21}M_{13} + P_{22}M_{23} \\ P_{32}M_{21} + P_{33}M_{31} & P_{31}M_{12} + P_{33}M_{32} & P_{31}M_{13} + P_{32}M_{23} \end{pmatrix}.$$

From Equation 3, we have

$$r_1 = P_{12}M_{21} + P_{13}M_{31} + 1$$

$$r_2 = P_{21}M_{12} + P_{23}M_{32} + 1$$

$$r_3 = P_{31}M_{13} + P_{32}M_{23} + 1$$

and the difference in the diagonal entries of D and the diagonal entries of PM is 1. Thus, if we add 1 to each of the diagonal entries of PM, we get the diagonal entries of D.

From Equation 2, we have

$$M_{ij} = 1 + \sum_{k \neq j} P_{ik}M_{kj}, \quad i \neq j.$$

If we take $i = 2$ and $j = 3$, in the three-state case, this says

$$M_{23} = 1 + P_{21}M_{13} + P_{22}M_{23},$$

so

$$M_{23} = (PM)_{23} + 1.$$

In general, as we showed in Equation 2, if we compare the off-diagonal entries of M with the off-diagonal entries of PM, we have $M_{ij} = (PM)_{ij} + 1$.

Let C be the $n \times n$ matrix, all of whose entries are 1. Then

$$M_{ij} = (PM + C)_{ij} \quad \text{for } i \neq j$$

and

$$D_{ii} = (PM + C)_{ii}.$$

Now $M_{ii} = 0$ and $D_{ij} = 0$ if $i \neq j$, so we have

$$PM + C = M + D. \tag{4}$$

Recall that we have set $E[T_i] = r_i$. We restate the theorem with that notation.

Theorem 2.19

Let $\{X_n\}$ be an irreducible finite-state Markov chain with equilibrium state

$$\hat{\pi} = (\pi(1), \ldots, \pi(n)),$$

and let r_i be the mean recurrence time of the state i. Then

$$\pi(i) = \frac{1}{r_i}.$$

Proof

By Equation 4, we have

$$PM + C = M + D,$$

so

$$C = D + M - PM = D + (I - P)M.$$

Then

$$\hat{\pi}C = \hat{\pi}\left[D + (I - P)M\right] = \hat{\pi}D \quad \text{since } \hat{\pi} = \hat{\pi}P.$$

Now

$$\hat{\pi}C = (\pi(1), \ldots, \pi(n))\begin{pmatrix} 1 & \cdots & 1 \\ \vdots & & \vdots \\ 1 & \cdots & 1 \end{pmatrix}$$

$$= (\pi(1) + \cdots + \pi(n), \ldots, \pi(1) + \cdots + \pi(n)) = (1, \ldots, 1)$$

and

$$\hat{\pi}D = (\pi(1), \ldots, \pi(n))\begin{pmatrix} r_1 & 0 & \cdots & 0 \\ 0 & r_2 & \cdots & 0 \\ \vdots & \vdots & & \vdots \\ 0 & 0 & \cdots & r_n \end{pmatrix} = (\pi(1)r_1, \ldots, \pi(n)r_n),$$

so

$$\pi(i)r_i = 1, i = 1, \ldots, n.$$

Fundamental Matrix for Regular Markov Chains

We showed earlier how to find the mean passage time between two states for an irreducible chain by converting the chain to an absorbing chain. While this method works, it gives only the mean first passage time to one particular state for each conversion and is inefficient if we want the mean first passage time between every pair of states. It may appear that Equation 4 of the previous section provides a way to find the matrix M (which is what we would like). It would, if $(I - P)^{-1}$ exists, but unfortunately, that is never the case as we show in Exercise 2.17. We next show how to find an explicit formula for M in the case of a regular Markov chain.

We saw that for absorbing Markov chains, the fundamental matrix

$$\boldsymbol{N} = (\boldsymbol{I} - \boldsymbol{Q})^{-1}$$

facilitated the computation of several quantities. For regular Markov chains, there is an analogous matrix also called the fundamental matrix, and it is denoted by $\boldsymbol{Z}$ in the literature.

For absorbing chains, each state is either transient or absorbing, and the matrix $\boldsymbol{Q}$ was made up of the transient (or non-absorbing) states. That we were able to define $\boldsymbol{N}$ by

$$\boldsymbol{N} = \boldsymbol{I} + \boldsymbol{Q} + \boldsymbol{Q}^2 + \boldsymbol{Q}^3 + \cdots$$

depended on the fact that

$$\lim_{n \to \infty} \boldsymbol{Q}^n = \boldsymbol{0}.$$

For a regular Markov chain, we need to find a matrix that will play a role that is analogous to $\boldsymbol{Q}$. If P is the transition matrix with limiting matrix A, then neither P^n nor A^n converges to 0. However,

$$\lim_{n \to \infty} P^n = A = A^n$$

and we might try using the matrix $(P - A)$ in a manner similar to what we did with the matrix $\boldsymbol{Q}$ in absorbing Markov chains. We give the properties of $P - A$ that we will need in the next theorem.

Theorem 2.20

Let P be the transition matrix of a finite-state regular Markov chain and let

$$A = \lim_{n\to\infty} P^n.$$

Then
(a) $PA = AP = A$.
(b) $(P - A)^n = P^n - A$.

Proof

We give the proof of part (b) and leave the proof of part (a) as an exercise.
Since $PA = AP$, we may use the Binomial Theorem to get

$$(P-A)^n = \sum_{k=0}^{n} \binom{n}{k} P^k (-A)^{n-k} = \sum_{k=0}^{n} \binom{n}{k} P^k (-1)^{n-k} (A)^{n-k}$$

$$= \sum_{k=0}^{n-1} \binom{n}{k} (-1)^{n-k} P^k (A)^{n-k} + P^n = A\sum_{k=0}^{n-1} \binom{n}{k} (-1)^{n-k} + P^n,$$

since if $k = n$, then $P^k(A)^{n-k} = P^n(A)^{n-n} = P^n$ and if $k < n$ then $P^k(A)^{n-k} = A$.

By the Binomial Theorem

$$0 = (1-1)^n = \sum_{k=0}^{n} \binom{n}{k} (1)^k (-1)^{n-k} = \sum_{k=0}^{n-1} \binom{n}{k} (1)^k (-1)^{n-k} + 1,$$

so

$$\sum_{k=0}^{n-1} \binom{n}{k} (1)^k (-1)^{n-k} = -1$$

and thus,

$$A\sum_{k=0}^{n-1} \binom{n}{k} (-1)^{n-k} + P^n = P^n - A.$$

Since

$$A = \lim_{n\to\infty} P^n \text{ and } P^n - A = (P-A)^n,$$

we conclude

$$\lim_{n\to\infty}(P-A)^n \text{ is the zero matrix.}$$

The previous theorem and the next few results deal with the relationships of certain matrices. We illustrate these with the example

$$P=\begin{pmatrix} .3 & .4 & .3 \\ .2 & .6 & .2 \\ .3 & .2 & .5 \end{pmatrix}.$$

All computations are done using a CAS.

We find A by raising P to a high power:

$$P^{100}\approx\begin{pmatrix} .2580 & .4193 & .3225 \\ .2580 & .4193 & .3225 \\ .2580 & .4193 & .3225 \end{pmatrix},$$

which we will take for A.

We find

$$PA=\begin{pmatrix} .3 & .4 & .3 \\ .2 & .6 & .2 \\ .3 & .2 & .5 \end{pmatrix}\begin{pmatrix} .2580 & .4193 & .3225 \\ .2580 & .4193 & .3225 \\ .2580 & .4193 & .3225 \end{pmatrix}$$

$$=\begin{pmatrix} .2580 & .4193 & .3225 \\ .2580 & .4193 & .3225 \\ .2580 & .4193 & .3225 \end{pmatrix}=A$$

and

$$AP=\begin{pmatrix} .2580 & .4193 & .3225 \\ .2580 & .4193 & .3225 \\ .2580 & .4193 & .3225 \end{pmatrix}\begin{pmatrix} .3 & .4 & .3 \\ .2 & .6 & .2 \\ .3 & .2 & .5 \end{pmatrix}$$

$$=\begin{pmatrix} .2580 & .4193 & .3225 \\ .2580 & .4193 & .3225 \\ .2580 & .4193 & .3225 \end{pmatrix}=A.$$

Also

$$P - A = \begin{pmatrix} .04193 & -.01935 & -.02258 \\ -.05806 & .1806 & -.1225 \\ .04193 & -.2103 & .1774 \end{pmatrix}$$

$$P^{10} - A = \begin{pmatrix} .3 & .4 & .3 \\ .2 & .6 & .2 \\ .3 & .2 & .5 \end{pmatrix}^{10} - \begin{pmatrix} .2580 & .4193 & .3225 \\ .2580 & .4193 & .3225 \\ .2580 & .4193 & .3225 \end{pmatrix}$$

$$= \begin{pmatrix} -1.613 \cdot 10^{-7} & 3.965 \cdot 10^{-7} & -2.803 \cdot 10^{-7} \\ -3.268 \cdot 10^{-6} & 1.116 \cdot 10^{-5} & -7.890 \cdot 10^{-6} \\ 4.341 \cdot 10^{-6} & -1.482 \cdot 10^{-5} & 1.048 \cdot 10^{-5} \end{pmatrix}$$

$$(P - A)^{10} = \begin{pmatrix} .04193 & -.01935 & -.02258 \\ -.05806 & .1806 & -.1225 \\ .04193 & -.2103 & .1774 \end{pmatrix}^{10}$$

$$= \begin{pmatrix} -1.613 \cdot 10^{-7} & 3.965 \cdot 10^{-7} & -2.803 \cdot 10^{-7} \\ -3.268 \cdot 10^{-6} & 1.116 \cdot 10^{-5} & -7.890 \cdot 10^{-6} \\ 4.341 \cdot 10^{-6} & -1.482 \cdot 10^{-5} & 1.048 \cdot 10^{-5} \end{pmatrix}$$

which tends to corroborate

$$P^n - A = (P - A)^n$$

and

$$\lim_{n\to\infty} (P - A)^n \text{ is the zero matrix.}$$

Theorem 2.21

The matrix Z defined by

$$Z = [I - (P - A)]^{-1}$$

exists and

$$Z = I + \sum_{n=1}^{\infty} \left(P^n - A\right).$$

Proof
Since

$$\lim_{n\to\infty} \left(P - A\right)^n \text{ is the zero matrix,}$$

we have

$$\left[I - \left(P - A\right)\right]^{-1} = I + \left(P - A\right) + \left(P - A\right)^2 + \left(P - A\right)^3 + \cdots$$

$$= I + \sum_{n=1}^{\infty} \left(P - A\right)^n = I + \sum_{n=1}^{\infty} \left(P^n - A\right).$$

For the matrices

$$P = \begin{pmatrix} .3 & .4 & .3 \\ .2 & .6 & .2 \\ .3 & .2 & .5 \end{pmatrix} \quad \text{and} \quad A = \begin{pmatrix} .2580 & .4193 & .3225 \\ .2580 & .4193 & .3225 \\ .2580 & .4193 & .3225 \end{pmatrix},$$

the matrix Z is

$$\left[I - \left(P - A\right)\right]^{-1}$$

$$= \left[\begin{pmatrix} 1 & 0 & 0 \\ 0 & 1 & 0 \\ 0 & 0 & 1 \end{pmatrix} - \begin{pmatrix} .04193 & -.01935 & -.02258 \\ -.05806 & .1806 & -.1225 \\ .04193 & -.2103 & .1774 \end{pmatrix}\right]^{-1}$$

$$= \begin{pmatrix} 1.4037 & -.0177 & -.0260 \\ -.0853 & 1.2726 & -.1873 \\ .0760 & -.3403 & 1.2643 \end{pmatrix}.$$

Theorem 2.22

Let P, A, M, C, D, and Z be defined as earlier. Then

(a) $I - Z = A - PZ$
(b) $ZC = C$
(c) $M = C - ZD + AM$

Proof

(a) We show $I - A = Z(I - P)$. Now $Z(I - P) = (I - P)Z$, so

$$Z(I-P)=(I-P)\left[I+\sum_{n=1}^{\infty}(P^n-A)\right]$$

$$=(I-P)+\sum_{n=1}^{\infty}(P^n-A)-P\sum_{n=1}^{\infty}(P^n-A)$$

$$=(I-P)+\sum_{n=1}^{\infty}(P^n-A)-\sum_{n=2}^{\infty}(P^n-A)=(I-P)+(P-A)=I-A.$$

(b) Let

$$\hat{c}=\begin{pmatrix}1\\ \vdots\\ 1\end{pmatrix}.$$

Then $I\hat{c}=\hat{c}$, $P\hat{c}=\hat{c}$ and $A\hat{c}=\hat{c}$, so

$$\hat{c}=(I-P+A)\hat{c}\quad\text{and thus }(I-P+A)^{-1}\hat{c}=\hat{c}\text{ so }Z\hat{c}=\hat{c}.$$

Then

$$ZC=Z(\hat{c},\ldots,\hat{c})=(Z\hat{c},\ldots,Z\hat{c})=(\hat{c},\ldots,\hat{c})=C.$$

(c) From Equation 4 of the previous section, we have

$$C-D=M-PM=M(I-P).$$

Now $M(I - P) = (I - P)M$ as we show in Exercise 2.32.

So

$$Z(C-D)=Z(I-P)M$$

and

$$ZC=C,Z(I-P)=I-A.$$

So

$$C-ZD=(I-A)M=M-AM$$

and thus

$$M=C-ZD+AM. \tag{5}$$

We could find M from Equation 5 if $(I-A)$ was invertible, but that is not the case. However, we do now have sufficient information to compute M.

Theorem 2.23

For a finite-state regular Markov chain, the mean first passage time from state i to state j, $i \neq j$ is

$$M_{ij}=\frac{Z_{jj}-Z_{ij}}{\pi_j},$$

where $\hat{\pi}=(\pi_1,\ldots,\pi_n)$ is the equilibrium state for P.

Proof

We have $C_{ij}=1$,

$$AM=\begin{pmatrix}\pi_1 & \cdots & \pi_n\\ \vdots & & \vdots\\ \pi_1 & \cdots & \pi_n\end{pmatrix}\begin{pmatrix}0 & M_{12} & M_{13} & \cdots & M_{1n}\\ M_{21} & 0 & M_{23} & \cdots & M_{2n}\\ \vdots & \vdots & \vdots & & \vdots\\ M_{n1} & \cdots & M_{(n-1)} & & 0\end{pmatrix}$$

so

$$(AM)_{ij}=\pi_1M_{1j}+\cdots+\pi_nM_{nj},\quad \text{where } M_{jj}=0.$$

Also

$$ZD = \begin{pmatrix} Z_{11} & \cdots & Z_{1n} \\ \vdots & & \vdots \\ Z_{n1} & \cdots & Z_{nn} \end{pmatrix} \begin{pmatrix} r_1 & 0 & \cdots & 0 \\ 0 & r_2 & \cdots & 0 \\ \vdots & \vdots & \cdots & \vdots \\ 0 & 0 & \cdots & r_n \end{pmatrix} = \begin{pmatrix} z_{11}r_1 & z_{12}r_2 & \cdots & z_{1n}r_n \\ z_{21}r_1 & z_{22}r_2 & \cdots & z_{2n}r_n \\ \vdots & \vdots & \cdots & \vdots \\ z_{n1}r_1 & z_{n1}r_2 & \cdots & z_{nn}r_n \end{pmatrix},$$

so

$$(ZD)_{ij} = Z_{ij}r_j.$$

Thus,

$$M_{ij} = C_{ij} - (ZD)_{ij} + (AM)_{ij}$$

gives

$$M_{ij} = 1 - M_{ij}r_j + \sum_k \pi_k M_{kj}. \tag{6}$$

Now $M_{jj} = 0$, so

$$0 = M_{jj} = 1 - Z_{jj}r_j + \sum_k \pi_k M_{kj}$$

and

$$\sum_k \pi_k M_{kj} = Z_{jj}r_j - 1.$$

Thus, Equation 5, $M = C - ZD + AM$, gives

$$M_{ij} = 1 - Z_{ij}r_j + \sum_k \pi_k M_{kj} = 1 - Z_{ij}r_j + Z_{jj}r_j - 1 = (Z_{jj} - Z_{ij})r_j = \frac{Z_{jj} - Z_{ij}}{\pi_j}.$$

Example

We conclude with the example where

$$P = \begin{pmatrix} .3 & .4 & .3 \\ .2 & .6 & .2 \\ .3 & .2 & .5 \end{pmatrix}.$$

From

$$A = \begin{pmatrix} .2580 & .4193 & .3225 \\ .2580 & .4193 & .3225 \\ .2580 & .4193 & .3225 \end{pmatrix},$$

we know π = (.258064, 4193548, .3225806) and we have computed

$$Z = \begin{pmatrix} 1.4037 & -.0177 & -.0260 \\ -.0853 & 1.2726 & -.1873 \\ .0760 & -.3403 & 1.2643 \end{pmatrix}.$$

We have

$$M_{ij} = \frac{Z_{jj} - Z_{ij}}{\pi_j}$$

so

$$M_{12} = \frac{Z_{22} - Z_{12}}{\pi_2} = \frac{1.2726 - (-.0177)}{.4193548} = 3.077$$

$$M_{13} = \frac{Z_{33} - Z_{13}}{\pi_3} = \frac{1.2643 - (-.0260)}{.3225806} = 4.000$$

$$M_{21} = \frac{Z_{11} - Z_{21}}{\pi_1} = \frac{1.4037 - (-.0853)}{.2580645} = 5.770$$

$$M_{23} = \frac{Z_{33} - Z_{23}}{\pi_3} = \frac{1.2643 - (-.1873)}{.3225806} = 4.500$$

$$M_{31} = \frac{Z_{11} - Z_{31}}{\pi_1} = \frac{1.4037 - .0760}{.2580645} = 5.145$$

$$M_{32} = \frac{Z_{22} - Z_{32}}{\pi_2} = \frac{1.2726 - (-.3403)}{.4193548} = 3.846.$$

Example

We return to the example

$$P = \begin{pmatrix} 1/2 & 0 & 1/2 \\ 1/3 & 1/3 & 1/3 \\ 1/4 & 1/4 & 1/2 \end{pmatrix}.$$

Then

$$A = \lim_{n \to \infty} P^n$$

and, in this case,

$$A \approx \begin{pmatrix} .353 & .176 & .471 \\ .353 & .176 & .471 \\ .353 & .176 & .471 \end{pmatrix}.$$

Now

$$Z = \left[I - (P - A) \right]^{-1} \approx \begin{pmatrix} 1.17 & -.24 & .07 \\ -.01 & 1.73 & -.17 \\ -.12 & .11 & 1.01 \end{pmatrix}.$$

Thus,

$$M_{21} = \frac{z_{11} - z_{21}}{\pi_1} \approx \frac{1.17 + .01}{.353} = 3.34$$

and

$$M_{31} = \frac{z_{11} - z_{31}}{\pi_1} \approx \frac{1.17 + .12}{.353} = 3.66,$$

which agrees (within round-off error) with the method of changing state 1 to an absorbing state.

Dividing a Markov Chain into Equivalence Classes

The material in this section is also applicable to the case of countably infinite-state Markov chains as well as finite-state Markov chains unless stated otherwise.

Definition

Let $\{X_n\}$ be a Markov chain with state space S. We say state j is accessible from state i if there is an $n \geq 0$ for which $P^n(i, j) > 0$. In this case, we write $i \rightarrow j$. We say that a set $C \subset S$ is closed if it is impossible to go from a state in C to a state not in C; said another way, if $i \in C$ and $j \in C^c$ then $P^n(i, j) = 0$ for every positive integer n.

Theorem 2.24

Let C be a closed set of a Markov chain with transition matrix P. Then the restriction of P to the states in C is a stochastic matrix.

The proof is left as Exercise 2.33.

Definition

Two states i and j of a Markov chain with transition matrix P are said to communicate with one another if it is possible to go from i to j and also from j to i (not necessarily in one step or even the same number of steps). This says there are $m, n \geq 0$ with $p_n(i, j) > 0$ and $p_m(i, j) > 0$.

We use the notation $i \leftrightarrow j$ if i and j communicate with one another.

Theorem 2.25

Communication of states is an equivalence relation.

Proof

We need to show that $\leftrightarrow$ is reflexive, that is $i \leftrightarrow i$; symmetric, that is if $i \leftrightarrow j$, then $j \leftrightarrow i$; and transitive, that is if $i \leftrightarrow j$ and $j \leftrightarrow k$, then $i \leftrightarrow k$.

The relation $\leftrightarrow$ is reflexive since for any state i, $p_0(i, i) = 1$

That $\leftrightarrow$ is symmetric is by definition.

For transitivity, suppose that $i \leftrightarrow j$ and $j \leftrightarrow k$. Then there are

$$m, n \geq 0 \text{ with}$$

$$p_m(i,j) > 0 \quad \text{and} \quad p_n(j,k) > 0.$$

So

$$p_{m+n}(i,k) = \sum_l p_m(i,l)\,p_n(l,k) \geq p_m(i,j)\,p_n(j,k) \quad (\text{taking } l = j)$$

and

$$p_m(i,j)\,p_n(j,k) > 0.$$

Corollary

An equivalence class of $\leftrightarrow$ consists of those states that communicate with one another. These classes partition the states of the Markov chain.

Proof

This follows from properties of equivalence relations.

Theorem 2.26

Let i be a recurrent state of the Markov chain $\{X_n\}$, and let $[i]$ denote the equivalence class of i with respect to $\leftrightarrow$. The set $[i]$ is a closed set.

Proof

Suppose that j is a state for which $i \to j$. Since i is a recurrent state, the process must return to the state i. Thus, $j \to i$ and so $i \leftrightarrow j$. So we have $j \in [i]$, and thus, $[i]$ is a closed set.

Definition

A Markov chain is said to be irreducible if the equivalence relation $\leftrightarrow$ has only one equivalence class.

This is equivalent to the previous definition of irreducible that said a Markov chain is irreducible if all states communicate with one another.

Another description of an irreducible Markov chain that is also common in the literature is an ergodic Markov chain.

Example

We repeat an earlier example to emphasize that an irreducible chain is not necessarily regular as the transition matrix

$$P=\begin{pmatrix}0 & 1\\ 1 & 0\end{pmatrix}$$

shows. With this matrix, all states communicate but

$$P^{2n}=\begin{pmatrix}1 & 0\\ 0 & 1\end{pmatrix} \text{ and } P^{2n+1}=\begin{pmatrix}0 & 1\\ 1 & 0\end{pmatrix}.$$

Theorem 2.27

Every state in a given equivalence class of the equivalence relation $\leftrightarrow$ is recurrent or every state in the equivalence class is transient.

Another way to state this result is that recurrence and transience of states are class properties of the equivalence relation $\leftrightarrow$

Proof

Suppose that i and j communicate and that i is recurrent. Then there are m and n with $p_m(i, j) > 0$ and $p_n(j, i) > 0$ and

$$\sum_r p_r(i,i)=\infty.$$

We will show

$$\sum_r p_{m+r+n}(j,j)=\infty.$$

As we have shown previously,

$$p_{m+r+n}(j,j)\geq p_m(i,j)p_n(j,i)p_r(i,i),$$

so

$$\sum_r p_{m+r+n}(j,j)\geq p_m(i,j)p_n(j,i)\sum_r p_r(i,i)=\infty$$

and so j is recurrent.

Thus, within an equivalence class, if one state is recurrent, then all are recurrent and so if one state is transient, then all are transient.

Corollary

In a finite-state irreducible Markov chain, all states are recurrent.

Theorem 2.28

Let $\{X_n\}$ be an irreducible Markov chain with finite-state space $\{1, \ldots, N\}$. Fix state i. Define the random variable

$$T_j = \min_n \{X_n = j | X_0 = i\}.$$

Then $E[T_j] < \infty$.

NOTE: This theorem is not true if the state space is infinite.

Proof

We begin by showing that if $i, j \in \{1, \ldots, N\}$, then $p_n(i, j) > 0$ for some n,

$$0 \le n \le N-1.$$

Fix states i and j. Since the chain is irreducible, it is possible to go from state i to the state j. The problem is to show this can be done with fewer than N steps. Consider a path from i to j. If this path has N or more steps, then some state must have been visited twice. If this is the case, there must have been a loop in the path, and there is a shorter path if we eliminate the loop.

Let

$$\delta_{i,j} = \max\{p_k(i,j) | 0 \le k \le N\} (\text{so } \delta_{i,j} > 0) \quad \text{and} \quad \delta = \min\{\delta_{i,j} | 1 \le i, j \le N\}.$$

Note that $\delta > 0$, since it is the minimum of a finite set of positive numbers.

Thus, we have for any states i and j the probability of starting at any state i and reaching state j sometime in the first N steps is at least δ. So

$$P(T_j \ge N) \le 1-\delta \quad \text{and} \quad P(T_j \ge kN) \le (1-\delta)^k$$

for k a positive integer. Since $P(T_j \ge m)$ decreases as m increases, if

$$kN \le m \le (k+1)N,$$

then

$$\sum_{m=kN}^{(k+1)N} P(T_j \ge m) \le N(1-\delta)^k.$$

Since T_j is a random variable whose range is contained in the positive integers, we have

$$E\left[T_j\right]=\sum_{n=1}^{\infty}P\left(T_j\geq n\right)=\sum_{k=1}^{\infty}\left(\sum_{m=kN}^{(k+1)N}P\left(T_j\geq m\right)\right)\leq N\sum_{k=1}^{\infty}\left(1-\delta\right)^k<\infty.$$

Periodic Markov Chains

Definition

State i is said to be periodic if there is a positive integer n for which

$$p_n\left(i,i\right)>0.$$

That is, if there is a positive probability that the process in state i will return to state i at a later time. The period of a periodic state i is the largest value of k that will divide every $n>0$ where $p_n(i,i)>0$.

Said another way, the period of a periodic state i is k, where

$$k=\gcd\left\{n>0 \text{ such that } P(X_n=i|X_0=i)>0\right\},$$

where gcd is an abbreviation for greatest common divisor. If $k=1$, the state is said to be aperiodic.

The *most agreeable* case for a Markov chain is to be aperiodic and irreducible, because then, in the finite case, there will be a single equilibrium state.

If the chain is reducible (i.e., not irreducible), then it can essentially be split into multiple chains, each of which is irreducible. The reducible Markov chain will have multiple equilibrium states that are formed by pasting together the equilibrium states of the multiple irreducible chains in different ways.

Having a single equilibrium state π with $\pi(i)>0$ is the ideal situation for a Markov chain, but being periodic of period $d>1$ will preclude this, as we show in this section.

Example

Consider the Markov chain whose digraph is given in Figure 2.6.

Here $p_4(0,0)>0$ and $p_6(0,0)>0$. Furthermore, the only way the process can return to 0 is to traverse each circle some integer number of times. It should seem plausible that $p_n(0,0)>0$ if and only if

$$n=4a+6b,\quad a,b\in\mathbb{Z},a,b\geq 0.$$

Any such n is divisible by 2, so 2 is the period of state i even though $p_2(i,i)=0$.

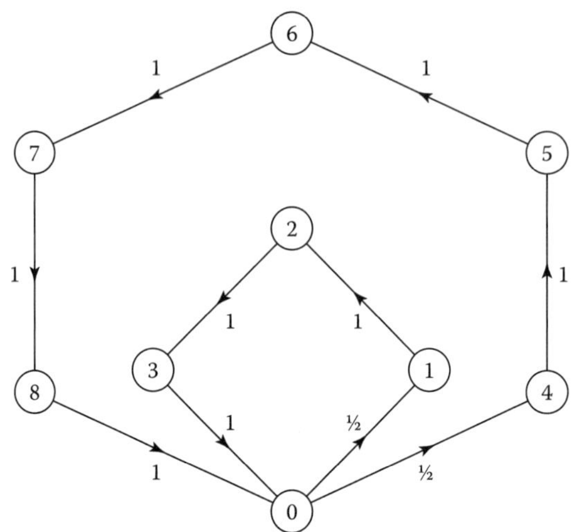

FIGURE 2.6
Digraph for Markov chain.

The problem here, as far as having the desired equilibrium situation is concerned, is that $p_{2k+1}(i, i) = 0$, so

$$\lim_{n\to\infty} p_n(i,i)$$

cannot exist and be nonzero.

This demonstrates that if the period of state i is $d > 1$, and if $\lim_{n\to\infty} p_n(i,i)$ exists, then that limit must be 0.

Example

If we modify the previous example slightly as shown in the digraph of Figure 2.7, we get very different results.

Now $p_n(0,0) > 0$ if and only if

$$n = 2a + 5b, \quad a,b \in \mathbb{Z}, a,b \geq 0.$$

Consider the possible values of such numbers n:

Multiples of 2 are 2, 4, 6, 8, ….

Multiples of 5 are 5, 10, 15, ….

Combinations of these multiples are 2, 4, 5, 6, 7, 8, 9, ….

It appears that all positive integers greater than or equal to 4 are in the last list. If this is the case, then $p_n(i, i) > 0$ if $n \geq 4$.

In fact, as soon as two consecutive integers appear in the list, we know state i is aperiodic (because the greatest common divisor of two consecutive integers is 1).

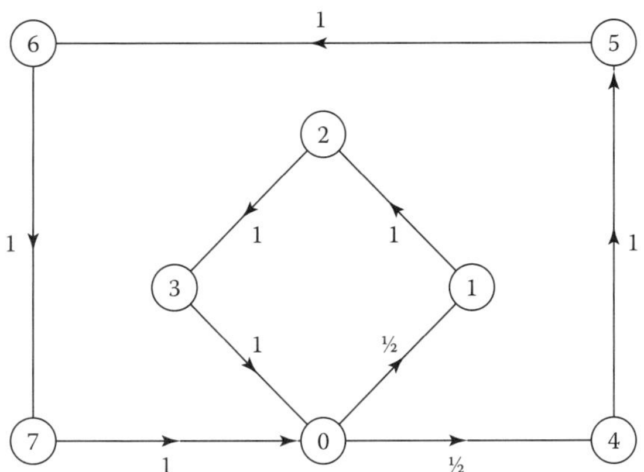

FIGURE 2.7
Digraph for Markov chain.

It is also the case that this example has a unique positive equilibrium state, as can be verified by taking a large power of the transition matrix. In fact, for this example, the transition matrix is

$$P = \begin{pmatrix} 0 & .5 & .5 & 0 & 0 & 0 \\ 1 & 0 & 0 & 0 & 0 & 0 \\ 0 & 0 & 0 & 1 & 0 & 0 \\ 0 & 0 & 0 & 0 & 1 & 0 \\ 0 & 0 & 0 & 0 & 0 & 1 \\ 1 & 0 & 0 & 0 & 0 & 0 \end{pmatrix}$$

and, using a CAS,

$$\lim_{n\to\infty} P^n = A,$$

where A is the matrix that has the identical rows

$$\left(\frac{2}{7},\frac{1}{7},\frac{1}{7},\frac{1}{7},\frac{1}{7},\frac{1}{7}\right).$$

We now proceed to prove several facts about periodicity. The most important of these are the following:

1. All states in the same communication class have the same period.
2. If a finite Markov chain is irreducible and aperiodic, then it is regular and, thus, has a unique equilibrium state.

In order to prove the first assertion, we need the following result:

Theorem 2.29

A set of positive integers that is closed under addition contains all but a finite number of multiples of its greatest common divisor.

Thus, if $p_n(i,i) > 0$ for two consecutive values of n, then $p_n(i,i) > 0$ for all n sufficiently large.

Proof

Let A be a nonempty set of positive integers and let d be the greatest common divisor of the terms of the set. If $d > 1$, then divide each element of A by d and call the resulting set A^*. If A^* has all but a finite number of positive integers, then A contains all but a finite number of multiples of its greatest common divisor.

A fact from number theory is that if d is the greatest common divisor of a set of integers B, then there are $b_1, \ldots, b_k \in B$ and integers $n_1, \ldots, n_k$ for which

$$d = n_1 b_1 + \cdots + n_k b_k.$$

Thus, there are $a_1, \ldots, a_k \in A^*$ and integers $n_1, \ldots, n_k$ for which

$$1 = n_1 a_1 + \cdots + n_k a_k.$$

Suppose that the terms are arranged so that $n_1, \ldots, n_j$ are positive integers and $n_{j+1}, \ldots, n_k$ are negative integers. Let

$$m = n_1 a_1 + \cdots + n_j a_j \quad \text{and} \quad -n = n_{j+1} a_{j+1} + \cdots + n_k a_k$$

so that

$$1 = m - n$$

and m and $-n$ are both in A^*.

Suppose that $s \geq n(n-1)$. By the fundamental theorem of arithmetic

$$s = qn + r \quad \text{where } 0 \leq r < n \text{ or } 0 \leq r \leq n-1. \quad \text{Also, } q \geq n-1,$$

so

$$s = qn + r = qn + r(m-n) = (q-r)n + rm.$$

Corollary

If A is a set of positive integers that is closed under addition and the greatest common divisor of the elements of A is 1, then there is a positive integer M for which $A \supset \{M, M+1, M+2, \ldots\}$.

Theorem 2.30

If states i and j communicate, then they have the same period.

Proof

Suppose states i and j communicate and let d_i be the period of state i and d_j be the period of state j.

Since i and j communicate, there are positive integers m and n with $p_m(i,j) > 0$ and $p_n(j,i) > 0$.

$$\text{Let } J_i = \{k \geq 0 \mid p_k(i,i) > 0\} \quad \text{and} \quad J_j = \{k \geq 0 \mid p_k(j,j) > 0\}.$$

Now

$$p_{m+n}(i,i) = \sum_k p_m(i,k)\,p_n(k,i) \geq p_m(i,j)\,p_n(j,i) > 0,$$

so $m+n \in J_i$. Likewise,

$$p_{n+m}(j,j) = \sum_k p_n(j,k)\,p_m(k,j) \geq p_n(j,i)\,p_m(i,j) > 0,$$

so $m+n \in J_j$. Thus,

$$d_i \mid m+n \quad \text{and} \quad d_j \mid m+n.$$

Now suppose

$$p_r(j,j) > 0; \quad \text{i.e., } r \in J_j.$$

We will show $r \in J_i$ by showing $d_i \mid r$. We have

$$p_{n+m+r}(i,i) = \sum_k p_n(i,k)\,p_n(k,j)\,p_r(j,j) \geq p_n(i,i)\,p_n(i,j)\,p_r(j,j) > 0,$$

so

$$n+m+r \in J_i; \quad \text{i.e., } d_i \big| m+n+r \quad \text{and} \quad \text{since } d_i \big| m+n \ \text{ then } \ d_i \big| r.$$

Thus, for any $r \in J_j$, d_i divides r. Since d_j is the greatest common divisor of the elements of J_j, we have $d_i \le d_j$.

A symmetric argument shows $d_j \le d_i$, so $d_i = d_j$.

Theorem 2.31

If a finite Markov chain is irreducible and aperiodic, then it is regular and thus has a single equilibrium state.

Proof

Let P be the transition matrix for a finite, irreducible, aperiodic Markov chain. As in the previous theorem, let

$$J_i = \{k \ge 0 | p_k(i,i) > 0\} \quad \text{and} \quad J_j = \{k \ge 0 | p_k(j,j) > 0\}.$$

Since the chain is aperiodic, for each state i, there is a positive integer $N(i)$ for which

$$J_i \supset \{N(i), N(i)+1, \ldots.\}.$$

Since P is irreducible and

$$p_{ij}^{\;r+n+m} \ge p_{ik}^{\;r} p_{kk}^{\;n} p_{kj}^{\;m}.$$

We have for each pair of states i, j there is a positive integer $M(i, j)$ such that

$$J_{ij} = \{k \ge 0 | p_k(i,j) > 0\} \supset \{M(i,j), M(i,j)+1, \ldots.\}.$$

Thus, P is regular.

We have seen that finite-state, irreducible, aperiodic Markov chains have a single equilibrium state. In this section and the next, we show how failing to have either of these characteristics can modify the situation. Earlier, we gave an intuitive argument that showed why a Markov with a periodic state cannot have a single equilibrium state. Here we expand our study of periodic Markov chains.

To guide our intuition for the abstract results, we will consider the Markov chain whose digraph is given in Figure 2.6. This is a Markov chain of period 2.

Suppose that $\{X_n\}$ is an irreducible Markov chain of period $d > 1$. Let $s \in S$ be a fixed state. For $k \in \{0, 1, \ldots, d-1\}$, define

$$A_k = \{x \in S \mid P^{nd+k}(s,x) > 0 \text{ for some nonnegative integer } n\}.$$

In the example, let 0 be the fixed state. There are two classes:

$$A_1 = \{x \mid P^{2n+1}(0,x) > 0 \text{ for some nonnegative integer } n\}$$

and

$$A_2 = \{x \mid P^{2n}(0,x) > 0 \text{ for some nonnegative integer } n\}$$

We determine A_1. Now

$$P(0,1) > 0,\ P(0,4) > 0,\ P(0,0) = 0 \text{ otherwise, so } \{1,4\} \subset A_1;$$

$$P^3(0,3) > 0,\ P^3(0,6) > 0,\ P^3(0,x) = 0 \text{ otherwise, so } \{3,6\} \subset A_1;$$

$$P^5(0,1) > 0,\ P^5(0,8) > 0,\ P^5(0,x) = 0 \text{ otherwise, so } \{1,8\} \subset A_1;$$

and if $P^{2n+1}(0,x) > 0$, then $x \in \{1,3,4,6,8\}$ so $A_1 = \{1,3,4,6,8\}$.

To determine A_2, we have

$$P^2(0,2) > 0,\ P^2(0,5) > 0,\quad P^2(0,x) = 0 \text{ otherwise, so } \{2,5\} \subset A_2;$$

$$P^4(0,0) > 0,\ P^4(0,7) > 0,\quad P^4(0,x) = 0 \text{ otherwise, so } \{0,7\} \subset A_2;$$

and if $P^{2n}(0,x) > 0$, then $x \in \{0,2,5,7\}$ so $A_2 = \{0,2,5,7\}$.

To explain the next idea, we will be slightly abusive of notation. We will say

$$P(a) = \{b,c\}$$

if $P(a,b) > 0$ and $P(a,c) > 0$ and $P(a,x) = 0$ otherwise. That is, b and c are exactly the states to which a can transition in one step.

Thus, we have

$$P(1) = \{2\},\ P(3) = \{0\},\ P(4) = \{5\},\ P(6) = \{7\},\ P(8) = \{0\}$$

so

$$P\{1,3,4,6,8\} = P\{A_1\} = \{0,2,5,7\} = A_2.$$

Similarly,

$$P(0) = \{1,4\}, \quad P(2) = \{3\}, \quad P(5) = \{6\}, \quad P(7) = \{8\}$$

so

$$P\{0,2,5,7\} = P\{A_2\} = \{1,3,4,6,8\} = A_1.$$

So

$$P : A_1 \to A_2 \quad \text{and} \quad P : A_2 \to A_1$$

and we have

$$P : A_k \to A_{k+1} \bmod 2.$$

For a periodic chain of period *d*, we have

$$P : A_k \to A_{k+1} \bmod d$$

and we have the cyclic diagram shown in Figure 2.8.

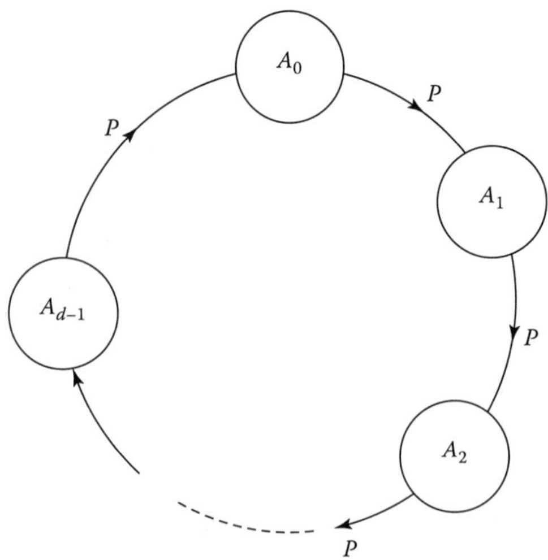

FIGURE 2.8
Evolution of the classes of a periodic chain of period *d*.

Figure 2.8 suggests that $x \in A_j$ and $y \in A_k$ if and only if $P^n(x, y) > 0$ for some n with

$$n \equiv (k - j) \bmod d$$

as we prove in Theorem 2.32.

Theorem 2.32

Let $y, z \in S$ with $y \in A_j, z \in A_k$. We have $P^n(y, z) > 0$ if and only if

$$n \equiv (k - j) \bmod d.$$

Proof

Since $\{X_n\}$ is irreducible, for any state y, there are positive integers m and n, for which

$$P^n(s, y) > 0 \quad \text{and} \quad P^m(y, s) > 0.$$

Then

$$P^{n+m}(s, s) \geq P^n(s, y) P^m(y, s) > 0,$$

so d divides $n + m$, and $n + m = qd$ for some integer q.

Now

$$P^n(s, y) > 0 \quad \text{and} \quad y \in A_j \text{ means } n = q_1 d + j.$$

So $n + m = qd$ and $n = q_1 d + j$, and thus,

$$qd = n + m = q_1 d + j + m \quad \text{and} \quad m = qd - (q_1 d + j) = q_2 d - j.$$

Also, $z \in A_k$, so $P^{rd+k}(s,z) > 0$ for some nonnegative integer r.

Now,

$$P^{m+n}(y, z) \geq P^m(y, s) P^n(s, z)$$

and if we take $m = q_2d - j$ and $n = rd + k$, then we have

$$P^{m+n}(y,z) = P^{(q_2d-j)+(rd+k)}(y,z) = P^{q_2d-j}(y,s)P^{rd+k}(s,z) > 0,$$

where $m + n = (q_2d - j) + (rd + k) = (q_2 + r)d + (k - j)$; that is,

$$n \equiv (k-j) \bmod d.$$

Corollary

For a periodic Markov chain of period $d > 1$, the sets A_k form a partition of the state space.

Proof

The equivalence classes of the integers modulo d partition the integers. The proof of the theorem shows that the mapping

$$A_k \to [k]$$

is an isomorphism.

Example

For periodic matrices, there is not an invariant probability distribution in the sense that

$$\lim_{n\to\infty} P^n(j,i)$$

exists and the value is independent of j. We can, however, make a related statement based on the following theorem:

Theorem 2.33

If P is an irreducible stochastic matrix with period d, then P has d eigenvalues with modulus 1. These eigenvalues are the roots of

$$z^d = 1.$$

Furthermore, if $\hat{v}$ is any probability distribution, then there is a unique probability distribution $\hat{\pi}$ for which

$$\lim_{n\to\infty}\frac{1}{d}\left[\hat{v}P^{n+1}+\cdots+\hat{v}P^{n+d}\right]=\pi.$$

The proof of this theorem is beyond the scope of the text.

This interpretation of $\hat{\pi}$ is that it gives the long-range proportion of time that the process spends in each state.

Example

Consider the Markov chain whose transition matrix is

$$P=\begin{pmatrix} 0 & 1/3 & 0 & 2/3 \\ 2/3 & 0 & 1/3 & 0 \\ 0 & 2/3 & 0 & 1/3 \\ 1/3 & 0 & 2/3 & 0 \end{pmatrix}$$

and the characteristic polynomial (using a CAS) is

$$(\lambda-1)(\lambda+1)\left(\lambda-\frac{i}{3}\right)\left(\lambda+\frac{i}{3}\right).$$

Let $\hat{v}=(.1,.3,.4,.2)$. Then,

$$\lim_{n\to\infty}\frac{1}{2}\left[\hat{v}P^{n+1}+\hat{v}P^{n+2}\right]=\left(\frac{1}{4},\frac{1}{4},\frac{1}{4},\frac{1}{4}\right).$$

The following example taken from http://www.math.uah.edu/stat/ is more complex.

Example

Consider

$$P=\begin{pmatrix} 0 & 0 & 1/2 & 1/4 & 1/4 & 0 & 0 \\ 0 & 0 & 1/3 & 0 & 2/3 & 0 & 0 \\ 0 & 0 & 0 & 0 & 0 & 1/3 & 2/3 \\ 0 & 0 & 0 & 0 & 0 & 1/2 & 1/2 \\ 0 & 0 & 0 & 0 & 0 & 3/4 & 1/4 \\ 1/2 & 1/2 & 0 & 0 & 0 & 0 & 0 \\ 1/4 & 3/4 & 0 & 0 & 0 & 0 & 0 \end{pmatrix}.$$

Using a CAS, one can show that P is of period 3 and

$$P^3 = \begin{pmatrix} 71/192 & 121/192 & 0 & 0 & 0 & 0 & 0 \\ 29/72 & 43/72 & 0 & 0 & 0 & 0 & 0 \\ 0 & 0 & 7/18 & 1/12 & 19/36 & 0 & 0 \\ 0 & 0 & 19/48 & 3/32 & 49/96 & 0 & 0 \\ 0 & 0 & 13/32 & 7/64 & 31/64 & 0 & 0 \\ 0 & 0 & 0 & 0 & 0 & 157/288 & 131/288 \\ 0 & 0 & 0 & 0 & 0 & 37/64 & 27/64 \end{pmatrix}.$$

So the cyclic classes are {1, 2}, {3, 4, 5}, {6, 7}.

The eigenvalues are

$$0, 1, -\frac{\sqrt[3]{57}}{12}, -\frac{1}{2}-\frac{i\sqrt{3}}{2}, -\frac{1}{2}+\frac{i\sqrt{3}}{2}, \frac{\sqrt[3]{57}}{24}-\frac{i3^{5/6}\sqrt[3]{57}}{24}, \frac{\sqrt[3]{57}}{24}+\frac{i3^{5/6}\sqrt[3]{57}}{24}$$

with

$$z = 1, -\frac{1}{2}-\frac{i\sqrt{3}}{2}, -\frac{1}{2}+\frac{i\sqrt{3}}{2} \quad \text{satisfying } z^3 = 1.$$

Letting $\hat{v} = (.1, .2, 0, .3, .1, .1, .2)$

$$\lim_{n\to\infty} \frac{1}{3}\left[\hat{v}P^{n+1} + \hat{v}P^{n+2} + \hat{v}P^{n+3}\right] \approx (.130, .203, .133, .032, .168, .187, .147).$$

Reducible Markov Chains

If P is a reducible Markov chain, then it has more than one equivalence class. A class is either transitive or recurrent, and at least one of the classes must be recurrent if the state space is finite. Also, a process that is in a transient state will eventually evolve to a recurrent state, and will remain in the recurrent class into which it has initially evolved. Thus, for any transient state j and any state i, we have

$$\lim_{n\to\infty} P^n(i, j) = 0.$$

Let $R_1, \ldots, R_m$ denote the recurrent classes and $T_1, \ldots, T_n$ denote the transient classes. The only way

$$\lim_{n\to\infty} P^n(i,j) \neq 0$$

is if j is in one of the recurrent classes. Also, a state cannot switch recurrent classes, since they are closed. Let i be a state in the recurrent class R_k. Then if

$$\lim_{n\to\infty} P^n(i,j) \neq 0,$$

it must be that $j \in R_k$.

We repeat a previous result.

Theorem 2.34

Every state in a given equivalence class of the equivalence relation $\leftrightarrow$ is recurrent or every state in the equivalence class is transient.

Corollary

The states Markov chain can be written as a disjoint union

$$T \cup R_1 \cup R_2 \cup \cdots,$$

where

T consists of the transient states

R_i are the distinct equivalence classes of recurrent states

Thus, in a reducible Markov chain, there is more than one equivalence class, and an equivalence class is either transient or recurrent and in a finite-state Markov chain, there is at least one recurrent class. In the long run, the process eventually ends up in a recurrent class, so we can consider the transient classes to be irrelevant as far as long-range behavior is concerned.

Each recurrent equivalence class is closed and can be considered a mini-Markov chain. As such, it is irreducible and if it is aperiodic, then it will have a unique equilibrium state. That equilibrium state can be extended to be an equilibrium state for the entire Markov chain by assigning 0 to each state outside the recurrent equivalence class.

Theorem 2.35

Every equilibrium state of a reducible aperiodic Markov chain is of the form

$$\sum a_i \pi^{(i)},$$

where $\pi^{(i)}$ is the unique equilibrium state of the ith recurrent equivalence class (extended to the whole chain) and $a_i \geq 0, \sum a_i = 1$.

Example

Consider the Markov chain whose transition matrix P is

$$\begin{array}{ccccccccc|c}
1 & 2 & 3 & 4 & 5 & 6 & 7 & 8 & 9 & \\
0 & 1/2 & 0 & 0 & 0 & 1/2 & 0 & 0 & 0 & 1 \\
1/2 & 0 & 1/2 & 0 & 0 & 0 & 0 & 0 & 0 & 2 \\
0 & 0 & 0 & 1/3 & 2/3 & 0 & 0 & 0 & 0 & 3 \\
0 & 0 & 1/2 & 0 & 1/2 & 0 & 0 & 0 & 0 & 4 \\
0 & 0 & 1/3 & 2/3 & 0 & 0 & 0 & 0 & 0 & 5 \\
0 & 0 & 0 & 0 & 0 & 0 & 1/2 & 1/3 & 1/6 & 6 \\
0 & 0 & 0 & 0 & 0 & 1/4 & 1/4 & 1/4 & 1/4 & 7 \\
0 & 0 & 0 & 0 & 0 & 1/2 & 1/2 & 0 & 0 & 8 \\
0 & 0 & 0 & 0 & 0 & 1/3 & 1/3 & 1/3 & 0 & 9
\end{array}$$

and whose digraph is shown in Figure 2.9.

The transient states are {1, 2} and there are two recurrent classes, {3, 4, 5} and {6, 7, 8, 9}.

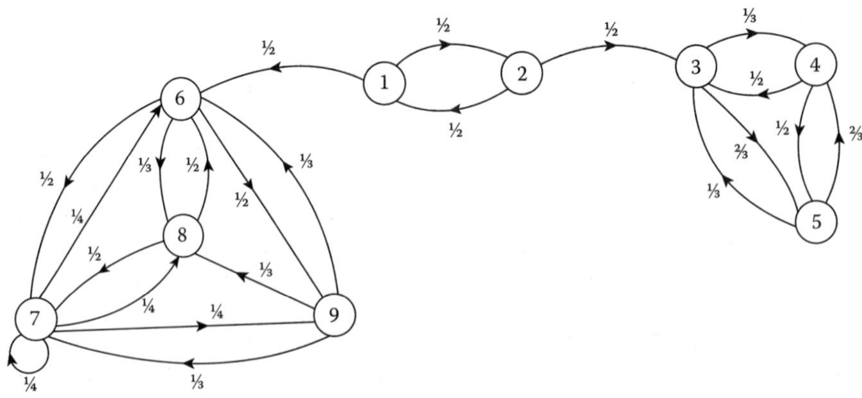

FIGURE 2.9
Digraph for Markov chain of the example.

To find the (approximate) equilibrium states for the two recurrent classes, we compute

$$P^{30} \approx \begin{pmatrix} 0 & 0 & .098 & .114 & .122 & .170 & .254 & .151 & .092 \\ 0 & 0 & .195 & .228 & .244 & .085 & .127 & .075 & .046 \\ 0 & 0 & .293 & .341 & .366 & 0 & 0 & 0 & 0 \\ 0 & 0 & .293 & .341 & .366 & 0 & 0 & 0 & 0 \\ 0 & 0 & .293 & .341 & .366 & 0 & 0 & 0 & 0 \\ 0 & 0 & 0 & 0 & 0 & .254 & .382 & .226 & .138 \\ 0 & 0 & 0 & 0 & 0 & .254 & .382 & .226 & .138 \\ 0 & 0 & 0 & 0 & 0 & .254 & .382 & .226 & .138 \\ 0 & 0 & 0 & 0 & 0 & .254 & .382 & .226 & .138 \end{pmatrix}.$$

Thus, P restricted to the recurrent class {3, 4, 5} has equilibrium state (.293, .341, .366) and $\pi^{(1)} = (0, 0, .293, .341, .366, 0, 0, 0, 0)$.

Also, P restricted to the recurrent class {6, 7, 8, 9} has equilibrium state (.254, .382, .266, .138) and $\pi^{(2)} = (0, 0, 0, 0, 0, .254, .382, .266, .138)$.

One can check that

$$\pi^{(1)}P = \pi^{(1)} \quad \text{and} \quad \pi^{(2)}P = \pi^{(2)}.$$

We have also asserted that if $a_1, a_2 \geq 0$ and $a_1 + a_2 = 1$, then $a_1\pi^{(1)} + a_2\pi^{(2)}$ is an equilibrium state for P.

Taking $a_1 = 1/3$ and $a_2 = 2/3$, we get

$$a_1\pi^{(1)} + a_2\pi^{(2)} = (0, 0, .098, .114, .122, .169, .255, .155, .092)$$

and one can check that

$$(0, 0, .098, .114, .122, .169, .255, .155, .092)P = (0, 0, .098, .114, .122, .169, .255, .155, .092).$$

Summary

Here we recap the major results of the chapter. In the recap, we assume that the state space is finite.

A principal goal with Markov chains is to determine the long-range behavior of the system. The evolution of a Markov process is determined by the transition matrix P. If there are n states in a Markov chain, then we describe the probability that the process is in a given state by a probability vector $(v_1, v_2, \ldots, v_n)$ where v_i is the probability the system is in state i. In discrete-time Markov processes, if the process has probability distribution $(v_1, v_2, \ldots, v_n)$ at time j, then it has probability distribution $(v_1, v_2, \ldots, v_n)P^k$ at time $j + k$. Thus, if the initial distribution is $(v_1, v_2, \ldots, v_n)$, then the long-term distribution is

$$\lim_{k\to\infty}(v_1, v_2, \ldots, v_n)P^k.$$

The *best* situation is when there is a unique probability distribution $(\pi_1, \ldots, \pi_n)$, called an equilibrium state, for which

$$\lim_{k\to\infty}(v_1, v_2, \ldots, v_n)P^k = (\pi_1, \ldots, \pi_n)$$

for every initial distribution $(v_1, v_2, \ldots, v_n)$. One way this is sure to occur is if P is a positive matrix, that is, if $P_{ij} > 0$ for all i and j. Another way is if P is regular; that is, if there is a positive integer N for which P^N is positive.

If P is positive or regular, then one can find the equilibrium state by finding the probability vector $(x_1, x_2, \ldots, x_n)$ for which

$$(x_1, x_2, \ldots, x_n) = (x_1, x_2, \ldots, x_n)P.$$

If P is positive or regular, then

$$\lim_{k\to\infty} P^k = A,$$

where A is a matrix whose rows are identical. A row of A is also the equilibrium state for the process.

States in a Markov chain can be classified as recurrent or transitive.

State i is recurrent if whenever the process starts in the state i, it returns to state i with probability 1. This implies that if the process begins in the state i, then it returns to the state i infinitely often with probability 1. The state i is transient if whenever the process starts in the state i, then it returns to the state i with probability less than 1.

For a transient state i, the expected number of visits to the state i, having begun in the state i, is

$$E(N_i \mid X_0 = i) = \frac{1}{1 - f_{ii}},$$

where f_{ii} is the probability the process returns to the state i having begun in the state i and the expected number of visits to the state i having begun in the state j is

$$E(N_i \mid X_0 = j) = \frac{f_{ij}}{1 - f_{ii}},$$

where f_{ij} is the probability the process visits the state j having begun in the state i.

The state i is transient if and only if

$$E(N_i \mid X_0 = i) < \infty,$$

which is true if and only if

$$\sum_{n=1}^{\infty} p_n(i,i) < \infty.$$

From the last statement, it follows that the state i is recurrent if and only if

$$\sum_{n=1}^{\infty} p_n(i,i) = \infty.$$

If j is a transient state, then for any state i, $\lim_{n\to\infty} p_n(i,j) = 0$.

In a finite state Markov chain, not all states are transient.

An absorbing state is a state for which if the process enters the state, it never leaves the state. The state i is absorbing if and only if $P(i,i) = 1$. An absorbing Markov chain is one for which any state can move to an absorbing state (not necessarily in one step) with positive probability. In a finite state absorbing Markov chain, the probability that any state will eventually be absorbed is 1.

The mean recurrence time of a state i, which we denote r_i, in an irreducible Markov chain, is

$$r_i = E(T_n \mid X_0 = i), \quad \text{where } T_n = \min\{n \geq 1 \mid X_n = i\}.$$

The mean first passage time from a state i to a state j, $i \neq j$, denoted M_{ij}, is

$$E(T_n \mid X_0 = i), \quad \text{where } T_n = \min\{n \geq 1 \mid X_n = j\}.$$

In a finite irreducible Markov chain with a unique equilibrium state

$$\hat{\pi} = (\pi_1, \ldots, \pi_n),$$

we have

$$\pi_i = \frac{1}{r_i}.$$

We define an equivalence relation on the states of a Markov chain by i is equivalent to j if it is possible to go from j to i and possible to go from i to j. In this case, we say that i and j communicate with one another. The equivalence classes are formed by the states that communicate with each other.

In an equivalence class, each state is transitive or each state is recurrent.

The period of a state i is the largest value of k that divides every positive integer n for which $p_n(i, i) > 0$. If this value is 1, the state is aperiodic. An aperiodic Markov chain is one in which every state is aperiodic.

In an equivalence class, each state has the same period.

A finite, irreducible, aperiodic Markov chain has a unique equilibrium state. A Markov chain that fails to be irreducible or aperiodic does not have a unique equilibrium state.

If a Markov chain is not irreducible, then it can be divided into transient states and a collection of recurrent equivalence classes. The recurrent classes can be chosen so that each is irreducible. If the original chain is aperiodic, then each of the irreducible equivalence classes is a miniature irreducible, aperiodic Markov chain that has its own unique equilibrium state. These equilibrium states can be put together in different ways to form different equilibrium states for the original Markov chain.

If P is the transition matrix for an irreducible Markov chain of period $d > 1$, then P will have d eigenvalues of modulus 1. There is an equilibrium distribution $\hat{\pi}$ such that for any initial probability distribution $\hat{v}$,

$$\lim_{n\to\infty} \frac{1}{d}\left(\hat{v}\,P^{n+1} + \hat{v}\,P^{n+2} + \cdots + \hat{v}\,P^{n+1} \right) = \hat{\pi}.$$

The interpretation of $\hat{\pi}$ is that it gives the long-range proportion of time that the process spends in each state.

Exercises

2.1 (a) Construct the transition matrix for the Markov chain associated with the diagram in Figure 2.10.

(b) Give (i) the transient states, (ii) the recurrent states, and (iii) the absorbing states.

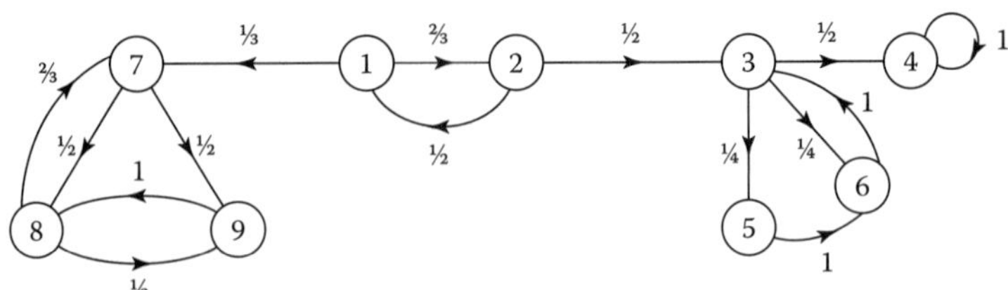

FIGURE 2.10
Digraph for Markov chain of Exercise 2.1.

(c) Is the chain reducible or irreducible? If it is reducible, give the equivalence classes.

(d) Give the period of each state.

2.2 For the Markov chain whose transition matrix is

$$P = \begin{pmatrix} .5 & .3 & .2 \\ .4 & .3 & .3 \\ .6 & .3 & .1 \end{pmatrix}.$$

(a) Find the equilibrium state by raising the matrix to a high power.

(b) Find the equilibrium state by solving $\pi P = \pi$.

(c) Find the expected return time for each state.

(d) Find the mean passage time between each pair of different states.

2.3 Repeat Problem 2 for the transition matrix

$$P = \begin{pmatrix} .1 & 0 & .3 & .6 \\ .2 & .4 & .3 & .1 \\ 0 & .3 & .5 & .2 \\ .1 & .4 & .1 & .4 \end{pmatrix}.$$

2.4 For the house in Figure 2.11, assume that a mouse moves from one room to another through a door of the room chosen at random.

(a) Construct the Markov chain associated with the diagram.

(b) Find the equilibrium state. How would you interpret the information given by the equilibrium state?

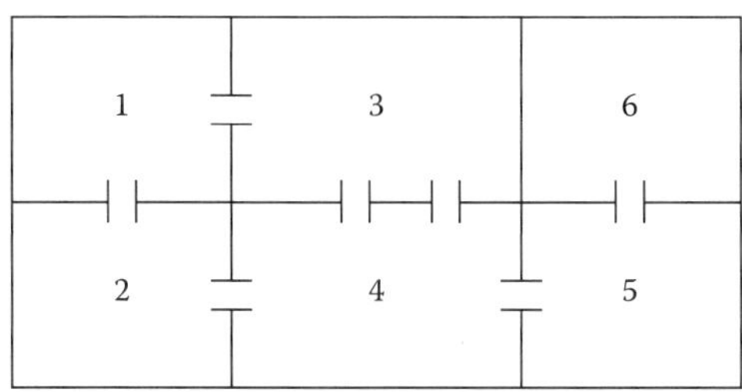

FIGURE 2.11
House for Exercise 2.4.

2.5 A substochastic A matrix is an $N \times N$ matrix for which $A_{ij} \geq 0$,

$$\sum_{j=1}^{N} A_{ij} \leq 1 \text{ and there is at least one i for which} \sum_{j=1}^{N} A_{ij} < 1.$$

In this exercise, we show any eigenvalue of an irreducible substochastic matrix has modulus less than 1.

(a) Give an example of a 2×2 substochastic matrix that has an eigenvalue equal to 1. Thus, irreducibility is a necessary condition.

(b) If A substochastic matrix, show that A^n is a substochastic matrix.

(c) Note that since A^n is a substochastic matrix $0 \leq r_i^n \leq 1$, and at least one $r_i^n < 1$. Let K be a row for which $r_K^1 \equiv r_K < 1$.

Show that

$$r_K^n = \sum_{j=1}^{N} A_{Kj} r_j^{n-1} \leq \sum_{j=1}^{N} A_{Kj} = r_K < 1.$$

(d) Show that for $m < N$ and any i, $1 \leq i \leq N$

$$r_i^N = \sum_{j=1}^{N} \left(M^m\right)_{ij} r_j^{N-m}$$

(e) Show that from part (d), it follows that A^N is positive and the sum of any row is less than 1.

(f) Let $r = \min_{1 \leq i \leq N} \left\{r_i^N\right\}$. So $r < 1$. Let

$$\hat{x} = \begin{pmatrix} x_1 \\ \vdots \\ x_N \end{pmatrix}$$

be a probability vector. Consider

$$A^N \begin{pmatrix} x_1 \\ \vdots \\ x_N \end{pmatrix}.$$

Show each entry of the vector is less than or equal to $r(x_1 + \cdots + x_N) = r < 1$. Conclude that any eigenvalue is less than 1.

2.6 Let P be the stochastic matrix

$$P = \begin{pmatrix} .5 & .3 & .2 \\ .1 & 0 & .9 \\ .4 & .3 & .3 \end{pmatrix}.$$

Use a CAS to find

(a) The eigenvalues of P (Answer: 1, 1/10, – 3/10)

(b) The equilibrium state by

 (i) finding P^{100}

 (ii) finding the left eigenvector for the eigenvalue 1

 $\left(\text{Answer} : \left(.3675, .2308, .4017\right)\right)$

2.7 For the diagrams in Figure 2.12:

(a) Give the transition matrix.

(b) Tell whether the associated chain is periodic.

(c) Classify each state as transient or recurrent.

2.8 For the transition matrix

$$P = \begin{pmatrix} .2 & .3 & .3 & .2 \\ .1 & .4 & .1 & .4 \\ .25 & .25 & .25 & .25 \\ .15 & .35 & .15 & .35 \end{pmatrix},$$

answer the following for the values in Theorem 2.6.

(a) What is ϵ?

(b) Compute $M - m$ for each column of P.

(c) Compute $M - m$ for each column of P^2.

(d) Compute $(1 - \epsilon)(M - m)$.

(e) Are these answers consistent with what is predicted by Theorem 2.6?

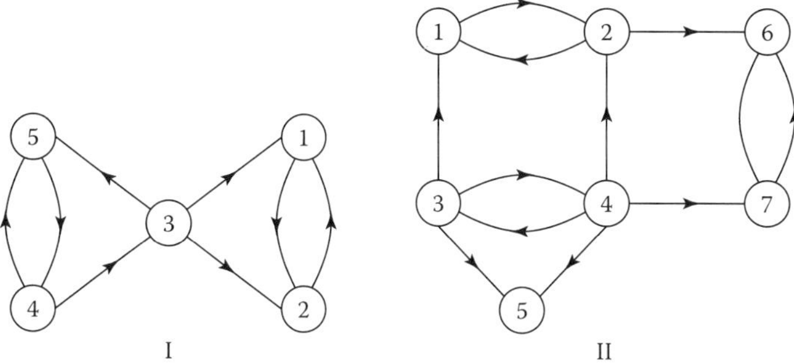

FIGURE 2.12
Digraph for Exercise 2.7.

(f) What is the equilibrium state?

(g) What is the expected return time of each state?

2.9 A matrix is doubly stochastic if both the matrix and its transpose are stochastic. Prove that a uniform distribution for a matrix P is a stationary distribution for P if and only if P is doubly stochastic.

2.10 There are three restaurants in a town, call them A, B, and C. A marketing survey gathered data that showed:

After eating at restaurant A, the next time a customer eats out 30% will return to A, 40% will go to B, and 30% will go to C.

After eating at restaurant B, the next time a customer eats out 20% will go to A, 30% will return to B, and 50% will go to C.

After eating at restaurant C, the next time a customer eats out 25% will go to A, 50% will go to B, and 25% will return to C.

(a) Construct the transition matrix when the aforementioned data is considered as a Markov chain.

(b) What is the predicted long-range share of the market for each restaurant?

(c) How many iterations are necessary before each state of the process will be within .01 of the steady state?

2.11 Alice, Brian, and Chloe are competitors in paintball. Alice hits her target 40% of the time, Brian hits his target 25% of the time, and Chloe hits her target 30% of the time.

The game is played so that all survivors shoot simultaneously, and each competitor shoots at the opponent that is most accurate.

Represent the game as a Markov chain where the states of the chain are the survivors.

(a) Give the transition matrix for the Markov chain.

(b) Which states are absorbing?

(c) What is the most likely outcome of the game?

(d) What is the expected length of the game?

2.12 A manufacturing company has a very long (theoretically infinite) chain of command. In passing an instruction, sometimes, it is not properly understood. Suppose that an instruction can be given by a *yes* or a *no*. Further suppose that when *yes* is given, it is correctly understood 95% of the time, and when *no* is given, it is correctly understood 90% of the time.

(a) Considering the process as a Markov chain, what are the states and the transition matrix?

(b) If a *yes* is given by the CEO, what is the probability the assembly line gets a *yes*?/ If a *no* is given by the CEO, what is the probability the assembly line gets a no?

2.13 We roll a dice until we get the string 663. Construct a Markov chain that will give the expected number of rolls to get the string 663.

(a) Give the transition matrix for the chain.

(b) Find the expected number of rolls to get 636.

2.14 Repeat Problem 12 for the string 6636.

2.15 (a) Give a transition matrix for a Markov chain that has 6 states and is irreducible and aperiodic.

(b) Give a transition matrix for a Markov chain that has 6 states and is irreducible and periodic of period 3.

(c) Give a transition matrix for a Markov chain that has 6 states and is reducible and aperiodic.

(d) Give a transition matrix for a Markov chain that has 6 states and is reducible and periodic of period 3.

2.16 For the example

$$P = \begin{pmatrix} .7 & .3 \\ .6 & .4 \end{pmatrix},$$

show that

$$E[T_2] = \frac{1}{\pi(2)}.$$

2.17 Show that for P a stochastic matrix, the matrix $(I - P)^{-1}$ does not exist.

2.18 A mouse lives in the house shown in Figure 2.13. He moves from room to room, selecting a door at random. So if a room has 3 doors, he will select each door with probability 1/3.

(a) Construct a Markov chain that describes the mouse's movements.

(b) In the long run, what percentage of time will the mouse spend in each room?

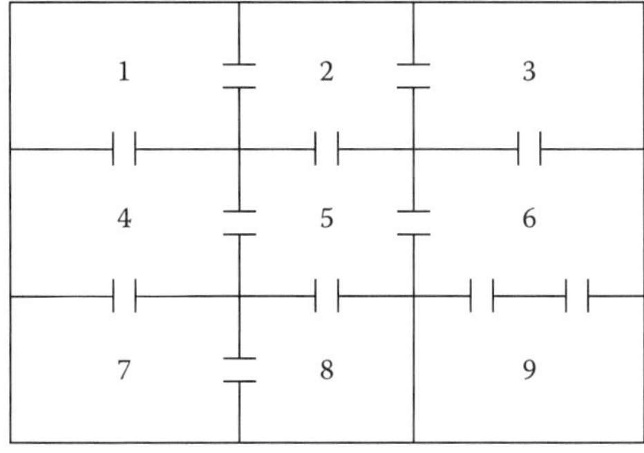

FIGURE 2.13
House for Exercise 2.18.

(c) If the mouse begins in room 3, what is the expected number of steps until he returns to room 3 for the first time?

(d) If the mouse begins in room 3, what is the expected number of steps until he returns to room 3 for the fifth time?

(e) If the mouse begins in room 3, what is the expected number of steps until he visits room 6 for the first time?

(f) If the mouse begins in room 3, what is the probability he visits room 6 before he visits room 1? (Hint: Revise the matrix so that rooms 6 and 1 are absorbing states.)

2.19 You are gambling against an opponent and between the two of you, you have \$10. You bet \$1 at a time. Are you more likely to be the eventual winner

(a) if you start with \$4 and your probability of winning an individual play of the game is .6, or if you start with \$6 and your probability of winning each play of the game is .4?

(b) if you start with \$3 and your probability of winning an individual play of the game is .7, or if you start with \$7 and your probability of winning an individual play of the game is .3?

(c) if you start with \$4, what should be the probability that you win an individual play of the game so that you and your opponent each have a probability of .5 of being the long-term winner?

2.20 Construct a Markov chain that has period 4. Give the graph and the transition matrix for the chain. Find the equilibrium state.

2.21 Construct a Markov chain that has some transition states and three equivalence classes. Give the graph and the transition matrix. What are the equivalence classes?

2.22 Show that a finite Markov chain can be partitioned into sets $T, C_1, \ldots, C_k$ so that T consists of transition states and the sets C_i are closed sets.

2.23 Show that a finite irreducible Markov chain is aperiodic if and only if for there is an n for which $(P^n)_{ij} > 0$ for all i and j.

2.24 Can the matrix below be the transition matrix for an irreducible periodic Markov chain? Explain your answer.

$$\begin{pmatrix} 0 & 0 & .5 & 0 & .2 & .3 \\ .1 & 0 & .6 & .2 & .1 & 0 \\ 0 & .2 & .1 & .6 & 0 & .1 \\ .4 & .2 & .1 & 0 & .3 & 0 \\ .8 & .2 & 0 & 0 & 0 & 0 \\ 1 & 0 & 0 & 0 & 0 & 0 \end{pmatrix}.$$

2.25 The transition matrix for a Markov chain is given in the following text. Find the absorbing states, the transient states, and the closed sets. (Number the states 1 through 7.)

$$\begin{pmatrix} 0 & .3 & 0 & .6 & .1 & 0 & 0 \\ 0 & 0 & 0 & .4 & 0 & .6 & 0 \\ 0 & 0 & 1 & 0 & 0 & 0 & 0 \\ .5 & .5 & 0 & 0 & 0 & 0 & 0 \\ 0 & .4 & 0 & .6 & 0 & 0 & 0 \\ 0 & 0 & 0 & 0 & 0 & 0.3 & .7 \\ 0 & 0 & 0 & 0 & 0 & .9 & .1 \end{pmatrix}.$$

2.26 Construct the transition matrix for the Ehrenfest model with 5 balls. What is the period? Find the equilibrium state. Conjecture what the equilibrium state would be for 7 balls. For $(2N + 1)$ balls.

2.27 We are stuck in Las Vegas, and are down to our last $20. Alas, a bus ticket home costs $80. We decide to try to get enough money for the ticket by gambling. The most favorable probability of winning a game that we have found is .4.

We consider two strategies. The more aggressive strategy is to bet all we have each time we play until we have $80 or go broke. A more conservative approach is to bet $5 each play of the game.

(a) What are the states of the Markov chain for each strategy?

(b) Give the transition matrix for each strategy.

(c) What is the probability of getting a bus ticket with each strategy?

(d) What is the expected number of trials until absorption with each strategy?

2.28 For the irreducible stochastic matrix

$$P = \begin{pmatrix} .1 & .4 & .3 & .2 \\ .2 & .2 & .5 & .1 \\ .6 & 0 & .2 & .2 \\ .3 & .5 & 0 & .2 \end{pmatrix}.$$

(a) Find the characteristic polynomial of P. Verify that 1 is an eigenvalue of P. (Such an eigenvector is assured by the Peron–Frobenius Theorem.)

(b) Find the eigenvector that is a probability vector associated with the eigenvalue 1.

(c) Use a CAS to find P^{10}. Compare the steady state predicted by both methods.

2.29 For the matrices below

(a) Give the transient and recurrent classes.

(b) Give the steady state for each of the recurrent classes.

(c) Give the general form of the equilibrium state.

$$P = \begin{pmatrix} .6 & .2 & .1 & 0 & .1 & 0 \\ .5 & .5 & 0 & 0 & 0 & 0 \\ 0 & 0 & .5 & .5 & 0 & 0 \\ 0 & 0 & .7 & .3 & 0 & 0 \\ 0 & 0 & 0 & 0 & .6 & .4 \\ 0 & 0 & 0 & 0 & .8 & .2 \end{pmatrix}$$

$$P = \begin{pmatrix} .8 & .2 & 0 & 0 & 0 & 0 \\ .5 & .5 & 0 & 0 & 0 & 0 \\ .1 & 0 & .3 & .6 & 0 & 0 \\ 0 & 0 & .7 & .3 & 0 & 0 \\ 0 & .2 & 0 & 0 & .4 & .4 \\ 0 & 0 & 0 & 0 & .8 & .2 \end{pmatrix}.$$

2.30 The matrix

$$P = \begin{pmatrix} 0 & 1 & 0 & 0 \\ 0 & 0 & 1 & 0 \\ 0 & 0 & 0 & 1 \\ 1 & 0 & 0 & 0 \end{pmatrix}$$

is the transition matrix for a Markov chain of period 4. Find the steady state π and show that for an probability vector $\hat{v}$

$$\lim_{n\to\infty} \frac{1}{4}\left[\hat{v}P^{n+1} + \hat{v}P^{n+2} + \hat{v}P^{n+3} + \hat{v}P^{n+4}\right] = \pi.$$

2.31 For the matrix

$$P = \begin{pmatrix} .5 & .3 & .2 \\ .4 & .4 & .2 \\ .3 & .3 & .4 \end{pmatrix},$$

find the matrices A, M, C, D, and Z of Theorem 2.20.

2.32 Show that in part (c) of Theorem 2.21,

$$M(I-P) = (I-P)M.$$

2.33 Show that if C is a closed set of a Markov chain with transition matrix P, then P restricted to C is a stochastic matrix.

3

Discrete-Time, Infinite-State Markov Chains

In this chapter, we study Markov chains whose state space is countably infinite. We will refer to these as countable chains. Many of the results from the finite-state case remain valid, but not all. One example of the differences is that the Markov chain can consist entirely of transient states as the following example shows.

Example

Suppose that the Markov chain $\{X_n\}$ has state space the positive integers and the transition probabilities are

$$P\{X_{n+1} = i+1 | X_n = i\} = 1.$$

Then all states are transitive.

Among the most significant differences between finite and countable Markov chains is determining the equilibrium state. We will see that for the eliminate-state case, we need to refine the idea of a recurrent state to a recurrent state being either null recurrent or positive recurrent. A crucial idea in the analysis of the existence and determination of an equilibrium state will be the expected length of time between the appearances of a state. We use renewal theory to determine some of the needed results, and this is our starting point for the present chapter.

Renewal Processes

We keep track of the times when some event occurs, say when a customer enters a store. Suppose these times are $\{t_n\}$ with

$$0 < t_1 < t_2 < \cdots.$$

We define a stochastic process $\{N(t)\}$ for $t \geq 0$ by

$N(t)$ = the number of times the event has occurred up to and including time t.

Thus, $N(t)$ is a random variable that takes values in the nonnegative integers.

An important observation is

$$N(t) = \begin{cases} n & \text{if } t_n \le t < t_{n+1}, n \ge 1 \\ 0 & \text{if } t < t_1 \end{cases}.$$

Let X_n, be the random variable that gives the time between the $(n-1)$st and nth occurrence of the event. So

$$X_n = \begin{cases} t_n - t_{n-1}, n > 1 \\ t_1, n = 1 \end{cases}$$

We call X_n an interarrival time. An important assumption is that the random variables $\{X_n\}$ are independent and identically distributed, and in this case, $\{N(t)\}$ is called a renewal process. When the event being counted takes place, in a probabilistic sense, the process is back at the starting point and we say a renewal has occurred.

Connecting these ideas with some results in Chapter 2, if i is a state of a countable Markov chain, we want to know frequently it appears. So we count the number of occurrences in time t (this is $N(t)$) and divide by the time, and this gives the relative frequency. We will use this as we did $E[T_i]$ in Chapter 2.

Example

For convenience, we rescale time so that changes occur only at time multiples of $t = .1$.

Suppose events occurred at $t = 1, 3.1, 4.5$, and 8. The graph of $N(t)$ versus t is shown in Figure 3.1.

Note that $N(t)$ takes values in the nonnegative integers and is continuous from the right. The values of X_n, $n = 1, 2, 3, 4$, in this example are $X_1 = 1, X_2 = 2.1, X_3 = 1.4, X_4 = 3.5$.

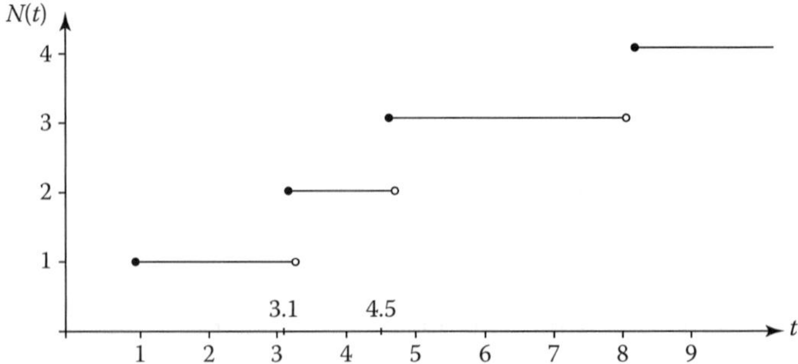

FIGURE 3.1
Graph of $N(t)$ versus t.

Note that

$$X_1 + X_2 + X_3 + \cdots + X_n = t_1 + (t_2 - t_1) + (t_3 - t_2) + \cdots + (t_n - t_{n-1}) = t_n.$$

Define the random variables S_n by

$$S_0 = 0, \quad S_n = \sum_{k=1}^{n} X_k, \, n \geq 1.$$

So

$$S_n = X_1 + X_2 + X_3 + \cdots + X_n = t_1 + (t_2 - t_1) + (t_3 - t_2) + \cdots + (t_n - t_{n-1}) = t_n$$

is the time of the nth occurrence of the event.

Since

$$N(t) = n \quad \text{if and only if } t_n \leq t < t_{n+1},$$

we have

$$N(t) = \max\{n | t_n \leq t\} = \max\{n | S_n \leq t\}. \tag{1}$$

We illustrate formula (1) with the example. There, we have

$$\begin{array}{ll} S_0 = 0 & N(t) = 0, \quad 0 \leq t < 1 \\ S_1 = X_1 = 1 & N(t) = 1, \quad 1 \leq t < 3.1 \\ S_2 = X_1 + X_2 = 3.1 & N(t) = 2, \quad 3.1 \leq t < 4.5 \\ S_3 = X_1 + X_2 + X_3 = 4.5 & N(t) = 3, \quad 4.5 \leq t < 8 \\ S_4 = X_1 + X_2 + X_3 + X_4 = 8 & N(t) = 4, \quad 8 \leq t. \end{array}$$

We have

$$t_n = \text{time of } n\text{th occurrence}$$

$$N(t) = \text{number of a occurrences up to time } t$$

$$S_n = t_n.$$

To determine $S_{N(t)}$:

1. Begin with t.
2. Find the largest value of n with $t_n \leq t$, i.e., how many occurrences have there been by time t. This is $N(t)$.
3. $S_{N(t)}$ is the time of the last occurrence prior to and including time t.

We consider the example. If $t = 3.2$, then the largest value of n with $t_n \le 3.2$ is 2. Thus, $N(3.2) = 2$, so

$$S_{N(3.2)} = S_2 = \text{time of second occurrence} = 3.1.$$

Thus, $N(t)$ is integer valued and $S_{N(t)}$ is the time of the most recent occurrence up to and including time t. Thus, $S_{N(t)}$ takes values in $(0, \infty)$ and

$$S_{N(t)} = t_{N(t)} \le t < t_{N(t)+1} = S_{N(t)+1}. \tag{2}$$

To illustrate (2), since the second occurrence takes place at time t_2 and the third occurrence takes place at time t_3, if $t_2 \le t < t_3$, then $N(t) = 2$ and $S_{N(t)} = t_2$.

In this case, we also have $N(t) + 1 = 3$, so

$$S_{N(t)+1} = t_{2+1} = t_3.$$

We return to the idea that S_n is the time of the nth occurrence of the event and if $t_n \le t < t_{n+1}$, then $S_n = t_n$. We thus have the relationship

$$N(t) \ge n \quad \text{if and only if } S_n \le t.$$

We will show that

$$EX_k = \mu > 0$$

if we assume that

$$P(X_k = 0) < 1.$$

The assumption means there is an $\varepsilon > 0$ and a $\delta > 0$ such that

$$P(X_k > \varepsilon) > \delta.$$

The fact that

$$EX_k = \mu > 0$$

with the assumption we have made follows from Markov's inequality, whose proof in the continuous case we now sketch.

Theorem 3.1 (Markov's Inequality)

Suppose that X is a nonnegative random variable and a is a number for which

$$P(X \ge a) > 0.$$

Then

$$E[X] \ge aP(X \ge a) > 0.$$

Proof

Since X is a nonnegative random variable, we have

$$E[X] = \int_0^\infty xf(x)dx \geq \int_a^\infty xf(x)dx \geq a\int_a^\infty f(x)dx = aP(X \geq a) > 0.$$

Corollary

If $P(X_k = 0) < 1$, then

$$EX_k = \mu > 0.$$

Proof

If we assume that

$$P(X_k = 0) < 1,$$

the assumption means there is an $\varepsilon > 0$ and a $\delta > 0$ such that

$$P(X_k > \varepsilon) > \delta.$$

The fact that

$$EX_k = \mu > 0$$

follows from Markov's inequality. Also, note that we have not precluded $\mu = \infty$.

We want to find expressions for

$$P\{N(t) = n\} \quad \text{and} \quad E[N(t)].$$

In Chapter 1, we showed that if X and Y are independent and identically distributed random variables that take values in the nonnegative integers with common density distribution f and cumulative distribution F, then $X + Y$ has the cumulative distribution

$$(F * f)(x) = \sum_{y=0}^{x} F(x-y)f(y).$$

We define $F_1 = F$ and

$$F_{k+1}(x) = \sum_{y=0}^{x} F_k(x-y)f(y).$$

Since the random variables X_k are independent and identically distributed, if F is the common cumulative distribution of each of the X_k's, then the distribution of

$$S_n = \sum_{k=1}^{n} X_k \quad \text{is } F_n.$$

NOTE: To ensure that X_n is integer valued, we require that changes occur at integer multiples of t. If necessary, this can be accomplished in the discrete case by a change of scale.

To find the distribution of $N(t)$, we use that $N(t) \geq n$ if and only if $S_n \leq t$. Thus,

$$P\{N(t)=n\} = P\{N(t) \geq n\} - P\{N(t) \geq n+1\} = P\{S_n \leq t\} - P\{S_{n+1} \leq t\}$$
$$= F_n(t) - F_{n+1}(t).$$

Since $N(t)$ takes values in the nonnegative integers, we have

$$E[N(t)] = \sum_{n=1}^{\infty} P[N(t) \geq n].$$

Now,

$$P[N(t) \geq n] = P[S_n \leq t] \quad \text{and} \quad P[S_n \leq t] = F_n(t),$$

so

$$E[N(t)] = \sum_{n=1}^{\infty} F_n(t).$$

Next, we prove two results that are crucial for renewal processes:

(i) For $t \in [0,\infty)$, $N(t) < \infty$,

(ii) $\lim_{t\to\infty} N(t) = \infty$.

Theorem 3.2

For $t \in [0, \infty)$, we have $N(t) < \infty$ with probability 1.

Proof

Fix $t_0 \in [0, \infty)$. Let $E[X_n] = \mu < \infty$. By the strong law of large numbers,

$$\lim_{n\to\infty} \frac{S_n}{n} = \lim_{n\to\infty} \frac{X_1 + \cdots + X_n}{n} = \mu$$

with probability 1. Because $\mu > 0$, as $n \to \infty$, we must have $S_n \to \infty$. Thus for any $\bar{t}$ there can be at most finitely many values of n with $S_n \le \bar{t}$; say $S_1, S_2, \ldots, S_N$.

Since

$$N(t) = \max\{n | S_n \le t\},$$

we have $N(\bar{t})$ is the largest $N(t)$ for which t does not exceed $\bar{t}$, and the result follows.

If $\mu = \infty$, the result is immediate.

The proof that

$$\lim_{t \to \infty} N(t) = \infty$$

is more involved. We begin with two definitions.

Definition

Let $E_1, E_2, E_3, \ldots$ be a sequence of events. We define

$$\limsup_n E_n = \cap_{n=1}^{\infty}\left(\cup_{k=n}^{\infty} E_k\right)$$

and

$$\liminf_n E_n = \cup_{n=1}^{\infty}\left(\cap_{k=n}^{\infty} E_k\right).$$

Theorem 3.3

We have

(a) $x \in \limsup_n E_n$ if and only if $x \in E_k$ for infinitely many sets E_k

(b) $x \in \liminf_n E_n$ if and only if $x \in E_k$ for all but finitely many sets E_k

(c) $\liminf_n E_n \subset \limsup_n E_n$

(d) $\left(\limsup_n E_n\right)^c = \liminf_n\left(E_n{}^c\right)$

The proof is left as Exercise 3.5.

Definition

A sequence of events $\{E_n\}$ is said to be increasing if $E_k \subset E_{k+1}$ for = 1, 2, ….

For an increasing sequence of events $\{E_n\}$, we define the event

$$\lim_{n \to \infty} E_n$$

by

$$\lim_{n\to\infty} E_n = \cup_{k=1}^{\infty} E_k.$$

A sequence of events $\{E_n\}$ is said to be decreasing if $E_k \supset E_{k+1}$ for = 1, 2, ….
For a decreasing sequence of events $\{E_n\}$, we define the event

$$\lim_{n\to\infty} E_n$$

by

$$\lim_{n\to\infty} E_n = \cap_{k=1}^{\infty} E_k.$$

Example

For $E_n = \left[0, 1 - \frac{1}{n}\right]$, $\{E_n\}$ is an increasing sequence of sets and

$$\lim_{n\to\infty} E_n = \cup_{k=1}^{\infty} E_k = [0,1).$$

For $E_n = [n, \infty)$, $\{E_n\}$ is a decreasing sequence of sets and

$$\lim_{n\to\infty} E_n = \cap_{k=1}^{\infty} E_k = \varnothing.$$

Theorem 3.4

(a) If $\{E_n\}$ is an increasing sequence of events, then

$$\lim_{n\to\infty} P(E_n) = P\left(\lim_{n\to\infty} E_n\right).$$

(b) If $\{E_n\}$ is a decreasing sequence of events, then

$$\lim_{n\to\infty} P(E_n) = P\left(\lim_{n\to\infty} E_n\right).$$

This theorem shows two continuity conditions of P as a set function.

Proof

(a) Suppose that $\{E_n\}$ is an increasing sequence of events. Define a sequence of events $\{F_n\}$ by

$$F_1 = E_1 \quad \text{and} \quad F_{k+1} = E_{k+1} \setminus E_k \quad \text{for } k = 1, 2, \ldots.$$

Then the collection of events $\{F_n\}$ is mutually exclusive, and for any $n \geq 1$, we have

$$\cup_{k=1}^{n} E_k = \cup_{k=1}^{n} F_k$$

and

$$\cup_{k=1}^{\infty} E_k = \cup_{k=1}^{\infty} F_k.$$

The series $\sum_{k=1}^{\infty} P(F_k)$ converges, since the sequence of partial sums is increasing and bounded above by 1. Thus,

$$\begin{aligned} P\left(\lim_{n\to\infty} E_n\right) &= P\left(\cup_{k=1}^{\infty} E_k\right) = P\left(\cup_{k=1}^{\infty} F_k\right) = \sum_{k=1}^{\infty} P(F_k) \\ &= \lim_{n\to\infty} \sum_{k=1}^{n} P(F_k) = \lim_{n\to\infty} P\left(\cup_{k=1}^{n} F_k\right) = \lim_{n\to\infty} P\left(\cup_{k=1}^{n} E_k\right) \\ &= \lim_{n\to\infty} P(E_n). \end{aligned}$$

(b) Suppose that $\{E_n\}$ is a decreasing sequence of events. Then $\left\{(E_n)^c\right\}$ is an increasing sequence of events. Now

$$P\left[(\cap E_n)^c\right] = 1 - P(\cap E_n),$$

and by De Morgan's laws,

$$(\cap E_n)^c = \cup\left(E_n{}^c\right).$$

Since $\left\{(E_n)^c\right\}$ is an increasing sequence of events, we may apply part (a) to conclude

$$P\left(\cup_{k=1}^{\infty} (E_k)^c\right) = \lim_{n\to\infty} P\left((E_n)^c\right).$$

Thus,

$$\lim_{n\to\infty} P\left((E_n)^c\right) = \lim_{n\to\infty}\left[1 - P(E_n)\right] = 1 - \lim_{n\to\infty}\left[P(E_n)\right].$$

Now,

$$P\left[\left(\cap_{k=1}^{\infty} E_k\right)^c\right] = 1 - P\left(\cap_{k=1}^{\infty} E_k\right) = 1 - \lim_{n\to\infty} P(E_n),$$

so

$$P\left(\lim_{n\to\infty} E_n\right) = P\left(\cap_{k=1}^{\infty} E_k\right) = \lim_{n\to\infty} P(E_n).$$

Corollary

Suppose that $A_1, A_2, A_3, \ldots$ is a collection of events. Then

$$P\left(\liminf_{n\to\infty} A_n\right) = \lim_{n\to\infty} P\left(\cap_{k=n}^{\infty} A_k\right).$$

Proof
Let

$$E_n = A_n \cap A_{n+1} \cap A_{n+2} \cap \cdots.$$

Thus, $E_n \subset E_{n+1}$, so $\{E_n\}$ is an increasing sequence of sets, and we have

$$P\left(\lim_{n\to\infty} E_n\right) = \lim_{n\to\infty} P(E_n).$$

Note that

$$\liminf_{n\to\infty} A_n = \cup_{n=1}^{\infty}\left(\cap_{k=n}^{\infty} A_k\right) = \cup_{n=1}^{\infty} E_n,$$

so

$$P\left(\liminf_{n\to\infty} A_n\right) = P\left(\cup_{n=1}^{\infty} E_n\right) = P\left(\lim_{n\to\infty} E_n\right) = \lim_{n\to\infty} P(E_n) = \lim_{n\to\infty} P\left(\cap_{k=n}^{\infty} A_k\right).$$

Theorem 3.5 (Second Borel–Cantelli Lemma)

If $A_1, A_2, \ldots$ is a sequence of independent events and if $\sum P(A_n) = \infty$, then

$$P\left(\limsup_n A_n\right) = 1.$$

Proof

Since $1 - x < e^{-x}$, we have

$$1 - P(A_n) < e^{-P(A_n)}.$$

By Theorem 3.3(d),

$$\left(\limsup_n A_n\right)^c = \liminf_n \left(A_n{}^c\right)$$

so

$$P\left[\left(\limsup_n A_n\right)^c\right] = P\left[\liminf_n \left(A_n{}^c\right)\right]$$

and by Theorem 3.4

$$P\left[\liminf_n \left(A_n{}^c\right)\right] = \lim_n P\left[\cap_{k=n}^{\infty}\left(A_k{}^c\right)\right].$$

Now

$$P\left[\cap_{k=1}^{\infty}\left(A_k{}^c\right)\right] = \prod_{k=1}^{\infty} P\left(A_k{}^c\right) = \prod_{k=1}^{\infty}\left[1 - P(A_k)\right] \le \prod_{k=1}^{\infty} \exp\left(-P(A_k)\right)$$

$$= \exp\left(-\sum_{k=1}^{\infty} P(A_k)\right) = 0.$$

So, since

$$P\left[\left(\limsup_n A_n\right)^c\right] = 0,$$

we have

$$P\left(\limsup_n A_n\right) = 1.$$

The next corollaries will be important in our study of renewal theory.

Corollary

Suppose is an event consisting of a set of outcomes to an experiment and $P(A > 0)$. Suppose also that the outcomes of different trials of the experiment are independent. If we perform infinitely many trials of the experiment, then the event occurs infinitely often with probability 1.

Proof

Let A_n denote the event A in the nth repetition of the experiment. Since

$$x \in \limsup_n A_n \quad \text{if and only if} \quad x \in A_n \quad \text{for infinitely many } n,$$

if $P(A_n) = \alpha > 0$, then

$$\sum_{n=1}^{\infty} P(A_n) = \infty,$$

and the result follows from the second Borel–Cantelli Lemma.

Corollary

Let S_n be the time of the nth occurrence of the event A. Then

$$\lim_{n\to\infty} S_n = \infty$$

with probability 1.

Proof

Since $P(X_n > 0) > 0$, there is a $t > 0$ for which $P(X_n > t) > 0$. Now,

$$S_n = X_1 + \cdots + X_n,$$

and the result follows from the previous corollary.

From this, we get the next result.

Corollary

We have with probability 1

$$\lim_{t\to\infty} N(t) = \infty.$$

The next two theorems will be important in finding the equilibrium state of countable Markov chains.

Theorem 3.6

With the notation given earlier, let $E[X_k] = \mu$. Then

$$\lim_{t\to\infty}\frac{N(t)}{t}=\frac{1}{\mu}$$

with probability 1.

Proof

We consider the case where $\mu < \infty$. The result is immediate if $\mu = \infty$.

We have

$$S_{N(t)} \le t < S_{N(t)+1},$$

so

$$\frac{S_{N(t)}}{N(t)} \le \frac{t}{N(t)} < \frac{S_{N(t)+1}}{N(t)}.$$

Now,

$$S_{N(t)} = X_1 + \cdots + X_{N(t)},$$

and

$$\lim_{t\to\infty} N(t) = \infty.$$

By the strong law of large numbers, if $\{X_n\}$ is a sequence of independent, identically distributed random variables with $E[X_k] = \mu$, then

$$\frac{X_1 + X_2 + \cdots + X_n}{n} \to \mu \text{ with probability 1.}$$

Thus,

$$\lim_{t\to\infty}\frac{S_{N(t)}}{N(t)} = \lim_{n\to\infty}\frac{X_1 + X_2 + \cdots + X_n}{n} = \mu \text{ with probability 1.}$$

Also,

$$\frac{S_{N(t)+1}}{N(t)} = \frac{S_{N(t)+1}}{(N(t)+1)}\frac{(N(t)+1)}{N(t)} \quad \text{so} \left[\lim_{t\to\infty}\frac{S_{N(t)+1}}{(N(t)+1)}\right]\left[\lim_{t\to\infty}\frac{(N(t)+1)}{N(t)}\right] = \mu$$

since

$$\lim_{t\to\infty}\frac{(N(t)+1)}{N(t)} = 1.$$

Thus,

$$\mu = \lim_{t\to\infty}\frac{S_{N(t)}}{N(t)} \le \lim_{t\to\infty}\frac{t}{N(t)} \le \lim_{t\to\infty}\frac{S_{N(t)+1}}{N(t)+1} = \mu.$$

Since

$$\lim_{t\to\infty}\frac{t}{N(t)} = \mu,$$

we have

$$\lim_{t\to\infty}\frac{N(t)}{t} = \frac{1}{\mu}.$$

Theorem 3.7 Elementary Renewal Theorem

We have

$$\lim_{t\to\infty}\frac{E[N(t)]}{t} \to \frac{1}{\mu}.$$

It may appear that this result should follow from the previous theorem, but the problem (in the language of real analysis) is that pointwise convergence of a sequence of functions does not necessarily mean convergence of the associated integrals. That is, one cannot conclude that

$$\lim_{n\to\infty}\int_a^b f_n(x)\,dx = \int_a^b \lim_{n\to\infty} f_n(x)\,dx$$

without some additional conditions.

The proof is beyond the scope of the text, but may be found at http://www.columbia.edu/~ks20/stochastic-I/stochastic-I-RRT.pdf.

Delayed Renewal Processes

As stated previously, we will use renewal theory to obtain results about the equilibrium state of countable Markov chains. In doing so, we will need to know the expected number of steps between two consecutive visits to a particular state that we denote j. If we know the process begins at state j, then we can use the results from renewal theory to determine this. If, however, the process begins in a state $i \neq j$, then we need to modify the argument somewhat. We assume the expected time until the first appearance until state j beginning in state i is finite.

The intuition is that if $N(t)$ is the number of times that state j occurs in the interval of time $[0, t)$ if the process begins in state j and $N^*(t)$ is the number of times that state j occurs in the interval until time $[0, t)$ if the process begins in state i, then it should be

$$\lim_{t\to\infty}\frac{N(t)}{t}=\lim_{t\to\infty}\frac{N^*(t)}{t}.$$

If the process begins in state i, we let it run until the first time it is in state j and after that point it is probabilistically the same as if the process had begun in state j. Said in another way, $N(t) + 1 \approx N^*(t)$ and for t sufficiently large,

$$\lim_{t\to\infty}\frac{N^*(t)}{t}=\lim_{t\to\infty}\frac{N(t)+1}{t}=\lim_{t\to\infty}\left(\frac{N(t)}{t}+\frac{1}{t}\right)=\lim_{t\to\infty}\frac{N(t)}{t}.$$

To be more rigorous, let X_1 be the time until the first occurrence of state j, given that the process begins in state i. For $n \geq 2$, let X_n be the time between the nth and $(n + 1)$st visit to state j. Then $\{X_2, X_3, \ldots\}$ are independent, identically distributed random variables. As before, we assume $E[X_n] = \mu$, $n \geq 2$ and let

$$S_n = X_1 + \sum_{k=2}^{n} X_k.$$

Then

$$\frac{S_n}{n}=\frac{X_1}{n}+\frac{1}{n}\sum_{k=2}^{n}X_k=\frac{X_1}{n}+\frac{n-1}{n}\frac{1}{n-1}\sum_{k=2}^{n}X_k,$$

so

$$\lim_{n\to\infty}\frac{S_n}{n}=\lim_{n\to\infty}\left(\frac{X_1}{n}+\frac{n-1}{n}\frac{1}{n-1}\sum_{k=2}^{n}X_k\right)$$

$$=\lim_{n\to\infty}\frac{X_1}{n}+\lim_{n\to\infty}\left(\frac{n-1}{n}\right)\lim_{n\to\infty}\left(\frac{1}{n-1}\sum_{k=2}^{n}X_k\right)=0+1\mu=\mu.$$

Thus, we have the analogs of the limit theorems for renewal processes.

Theorem 3.8

For a delayed renewal process with $E[X_n]=\mu$ for $n\geq 2$,

(a) $\lim_{t\to\infty}\frac{N(t)}{t}=\frac{1}{\mu}$ with probability 1.

(b) $\lim_{t\to\infty}\frac{E\left[N(t)\right]}{t}=\frac{1}{\mu}$.

Equilibrium State for Countable Markov Chains

In this section, we determine when an equilibrium state exists for a countable Markov chain and whether such a state is unique. We consider only the case of an irreducible chain and initially assume that the chain is aperiodic.

A major difference between finite state and countably infinite state Markov chains is that an aperiodic, irreducible countable Markov chain does not always have an equilibrium probability distribution.

The goal of this section is to find conditions when a countably infinite, aperiodic, irreducible Markov chain has an equilibrium distribution. In this chapter, we consider only the discrete-time case, where states change only at integer values of time.

The material on renewal processes will be indispensible. We will show that a state π that satisfies

$$\pi(i)=\frac{1}{\mu(i)},$$

where $\mu(i)$ is the expected time between occurrences of state i that was derived in the section on renewal processes, has the property that

$$\pi P=\pi.$$

The difficulty is that $\mu(i)$ may be infinite for all i, and in that case, $\pi=0$ and is not a probability distribution. We will see that for an irreducible chain if $\mu(i)$ is infinite for one state, then it is infinite for all states.

Physical Interpretation of the Equilibrium State

The equilibrium state of a process could be described by the fraction of time each state appears if we allow the process to run for a long period of time. If $\pi = (\pi(i))$ is the equilibrium state with this interpretation, then if we could determine the expected number of times state i appears in a given number of consecutive steps, we could compute $\pi(i)$ and thus compute π.

We have previously classified states as being transient or recurrent and shown that in a given equivalence class of states, every state is transitive or every state is recurrent. It will now be necessary to further divide recurrent states into being null recurrent or positive recurrent. We will show that in a given equivalence class, each state is transient, null recurrent, or positive recurrent. This was not necessary in the finite state case because there are no null recurrent states in that setting.

Null Recurrent versus Positive Recurrent States

Let $f_{ij}(n)$, $i \neq j$ be the probability that starting in state i, the first entry to state j occurs on the nth step. That is,

$$f_{ij}(n) = P\{X_n = j, X_{n-1} \neq j, X_{n-2} \neq j, \ldots, X_1 \neq j | X_0 = i\}.$$

If $i = j$, then by convention, n is the first step where the process returns to state i. (So $n \geq 1$ for both $i \neq j$ and $i = j$.) This is the same definition as with finite Markov chains.

We define

$$F_{ij}(n) = \sum_{k=1}^{n} f_{ij}(k).$$

So $F_{ij}(n)$ is the probability that starting in state i, the process has made its first visit to state j by time n. Now, $F_{ij}(n)$ is a nonnegative, nondecreasing function that is bounded above by 1. Then

$$F_{ij}(\infty) \equiv \lim_{n \to \infty} F_{ij}(n)$$

exists and is the probability the process will ever visit state j having begun in state i.

If

$$F_{ij}(\infty) = 1,$$

then, beginning in state i, the probability that the process will eventually visit state j is 1. If this is the case, then we can define a random variable T_{ij} whose value is the positive integer that is the time of the first visit to state j having begun in state i. The probability density function of T_{ij} is f_{ij} and the cumulative density function is F_{ij}. Also, we define T_i (rather than T_{ii}) to be the random variable whose value is the positive integer that is the time of the first return visit to state i having begun in state i. (This is in keeping with the notation common in the literature and the notation of Chapter 2.)

Consistent with our definition in the finite-state case, we say state i is recurrent if $F_{ii}(\infty) = 1$ and transient if $F_{ii}(\infty) < 1$.

States that are recurrent can be divided into classes called null recurrent states and positive recurrent states that we will define shortly. States in the different classes manifest different long-range behavior.

Let

$$T_i = \inf\{n \geq 1 \quad \text{for which } X_n = i | X_0 = i\}.$$

Thus, T_i is the time that the process first returns to state i, having begun in state i. Again, this is what was done in the finite-state case. In the aforementioned notation,

$$P(T_i = n) = f_{ii}(n),$$

which is the probability the process will first return to state i after n steps.

The mean recurrence time of a state i is

$$m_i = E[T_i] = \sum_{n=1}^{\infty} n f_{ii}(n),$$

so m_i is the expected number of steps between two visits to state i. It is possible for state i to be recurrent and $m_i = \infty$ and possible for state i to be recurrent and $m_i < \infty$. This did not occur in the finite-state case.

Definition

The recurrent state i is positive recurrent if m_i is finite and null recurrent if $m_i = \infty$. Equivalently, the recurrent state i is positive recurrent if $\mu(i) > 0$ and null recurrent if $\mu(i) = 0$ where

$$\mu(i) = \frac{1}{m_i}.$$

Table 3.1 gives the relations we have discussed.

We now list some criteria for determining the classification of states.

TABLE 3.1

Classification of States

State	$\sum f_{ii}(n)$	$\sum nf_{ii}(n) = E[T_i] = m_i$
Transient	∞	∞
Null recurrent	$<\infty$	∞
Positive recurrent	$<\infty$	$<\infty$

Theorem 3.9

An irreducible Markov chain is transient if and only if

$$\sum_{n=0}^{\infty} p_n(i,i) < \infty.$$

This is the same as in the finite case.

Theorem 3.10

Suppose that X_n is an irreducible Markov chain with state space S. Fix a state j. Define

$$\alpha(i) = P\{X_n = j \quad \text{for some } n \geq 0 | X_0 = i\}.$$

The Markov chain X_n is transient if and only if for any state j there is a unique solution to the following three conditions:

(a) $0 \leq \alpha(i) \leq 1$.
(b) $\alpha(j) = 1, \inf\{\alpha(i) \mid i \in S\} = 0$.
(c) For $i \neq j$, $\alpha(i) = \sum_{k \in S} p(i,k)\alpha(k)$.

The proof of this theorem is beyond the scope of the text. We will use it again in Chapter 5.

We state some facts about transience, null recurrence, and positive recurrence.

Recall that an equilibrium probability distribution $\pi = (\pi(1), \pi(2), \ldots)$ is a probability distribution for which

$$\lim_{n\to\infty} p_n(y,x) = \pi(x)$$

for every pair of states x and y.

Some handy facts are as follows:

1. If X_n is a transient Markov chain, then

$$\lim_{n\to\infty} p_n(y,x) = 0$$

for every pair of states x and y.

2. The Markov chain X_n is null recurrent if it is recurrent and

$$\lim_{n\to\infty} p_n(y,x) = 0$$

for every pair of states x and y.

3. Positive recurrence is a class property. This will be proven in Theorem 3.12.

The next theorem is a result of major importance. The proof follows from these statements.

Theorem 3.11

An infinite irreducible aperiodic Markov chain is positive recurrent if and only if there is a probability distribution π for which $\pi P = \pi$.

Theorem 3.12

Positive recurrence is a class property of the equivalence relation $\leftrightarrow$. That is, if one state in an equivalence class is positive recurrent, then all states are positive recurrent.

Proof

Suppose state x is positive recurrent. This occurs if and only if $\mu(x) > 0$.

Suppose x and y in the same equivalence class with respect to the equivalence relation $\leftrightarrow$. We want to show $\mu(y) > 0$.

Since x and y in the same equivalence class, there are positive integers m and l for which $P^m(x, y) > 0$ and $P^l(y, x) > 0$.

Now,

$$P^{l+k+m}(y,y)=\sum_z P^l(y,z)P^k(z,z)P^m(z,y)\geq P^l(y,x)P^k(x,x)P^m(x,y).$$

Summing over k from $k = 1$ to n, we have

$$\sum_{k=1}^{n}P^{l+k+m}(y,y)\geq P^l(y,x)\left[\sum_{k=1}^{n}P^k(x,x)\right]P^m(x,y)=K\left[\sum_{k=1}^{n}P^k(x,x)\right],$$

where $K = P^l(y, x)P^m(x, y)$.

We want to show

$$\mu(y)\geq K\mu(x)>0.$$

This would show that y is positive recurrent.

Now,

$$\sum_{k=1}^{n}P^{l+k+m}(y,y)=P^{l+1+m}(y,y)+\cdots+P^{l+n+m}(y,y)$$

$$=\left[P^1(y,y)+\cdots+P^{l+m}(y,y)+P^{l+1+m}(y,y)+\cdots+P^{l+n+m}(y,y)\right]$$

$$-\left[P^1(y,y)+\cdots+P^{l+m}(y,y)\right]$$

$$=\left[\sum_{j=1}^{l+m+n}P^j(y,y)\right]-\left[\sum_{j=1}^{l+m}P^j(y,y)\right],$$

so

$$\sum_{k=1}^{n}P^{l+k+m}(y,y)=\left[\sum_{j=1}^{l+m+n}P^j(y,y)\right]-\left[\sum_{j=1}^{l+m}P^j(y,y)\right]\geq K\left[\sum_{k=1}^{n}P^k(x,x)\right].$$

Then

$$\frac{1}{l+n+m}\left[\sum_{k=1}^{n}P^{l+k+m}(y,y)\right]=\frac{1}{l+n+m}\left[\sum_{j=1}^{l+m+n}P^j(y,y)\right]-\frac{1}{l+n+m}\left[\sum_{j=1}^{l+m}P^j(y,y)\right]$$

$$\geq K\frac{1}{l+n+m}\left[\sum_{k=1}^{n}P^k(x,x)\right]. \tag{1}$$

We want to take the limit as $n \to \infty$ in expression (1). By definition,

$$\lim_{n\to\infty} \frac{1}{l+n+m}\left[\sum_{j=1}^{l+m+n} P^j(y,y)\right] = \mu(y).$$

Since

$$\sum_{j=1}^{l+m} P^j(y,y)$$

is a finite sum,

$$\lim_{n\to\infty} \frac{1}{l+n+m}\left[\sum_{j=1}^{l+m} P^j(y,y)\right] = 0.$$

Also,

$$\frac{1}{l+n+m}\left[\sum_{k=1}^{n} P^k(x,x)\right] = \frac{n}{l+n+m}\cdot\left[\frac{1}{n}\sum_{k=1}^{n} P^k(x,x)\right],$$

and

$$\lim_{n\to\infty} \frac{n}{l+n+m} = 1, \quad \lim_{n\to\infty}\left[\frac{1}{n}\sum_{k=1}^{n} P^k(x,x)\right] = \mu(x).$$

Thus, we have

$$\lim_{n\to\infty}\left\{\frac{1}{l+n+m}\left[\sum_{j=1}^{l+m+n} P^j(y,y)\right] - \frac{1}{l+n+m}\left[\sum_{j=1}^{l+m} P^j(y,y)\right]\right\}$$

$$\geq \lim_{n\to\infty}\left\{K\frac{1}{l+n+m}\left[\sum_{k=1}^{n} P^k(x,x)\right]\right\},$$

so

$$\mu(y) - 0 = \mu(y) \geq K\mu(x) > 0$$

and y is positive recurrent.

Corollary

The properties of a state being transient, null recurrent, or positive recurrent are all class properties.

The next theorem states why the issue of positive recurrence did not come up in the finite-state case.

Theorem 3.13

A finite-state Markov chain has no null recurrent classes.

Proof

Suppose that i is a null recurrent state in a finite-state Markov chain. Then every state in the equivalence class of i is null recurrent. Let C denote this equivalence class. We have shown

$$\lim_{n\to\infty} p_n(i,j) = 0 \quad \text{for all } i,j \in C \quad \text{and} \quad \sum_{j\in C} p_n(i,j) = 1 \quad \text{for all } n.$$

This yields the contradiction

$$1 = \lim_{n\to\infty} \sum_{j\in C} p_n(i,j) = \sum_{j\in C} \lim_{n\to\infty} p_n(i,j) = \sum_{j\in C} 0 = 0,$$

where moving the limit inside the sum is legitimate because the sum is finite.

Corollary

In a finite-state irreducible Markov chain, all states are positive recurrent.

We continue to apply the results from renewal theory. Suppose that i and j are states that communicate and we want to determine the average number of times state j appears in a certain period of time, which is equivalent to finding the average number of times state j occurs per unit time. If $i = j$, then this can be seen as a renewal process, and if $i \neq j$, then this is a delayed renewal process.

In the "Renewal Processes" section, $N(t)$ was the number of times some event occurred in time t. We want to know how many times state i appears in time t, and we will denote this $N_i(t)$.

Recall that if i is a state and $N_i(t)$ is the number of times state i appears in time t, then with probability 1,

$$\lim_{t\to\infty} \frac{N_i(t)}{t} = \frac{1}{\mu(i)}$$

and

$$\lim_{t\to\infty}\frac{E[N_i(t)]}{t}=\frac{1}{\mu(i)}$$

for both renewal and delayed renewal processes. Here, $\mu(i)$ is the expected time between consecutive occurrences of state i.

Suppose we start the process in state i and we want to determine the expected number of steps until the process makes its first return to state i. One way to do this is to let the process run for a long period of time and count the number of times state i appears. If t is the length of time the process has run and $N_i(t)$ is the number of times state i has appeared, then

$$\frac{t}{N_i(t)}$$

gives the average number of steps between consecutive visits to state i.

If we let Y_n be the time the process makes the nth visit to state i (and $Y_0 = 0$), then

$$Z_n = Y_n - Y_{n-1}$$

is the time between the nth and $(n-1)$st visit to state i. Then $\{Z_n\}$ is a collection of random variables that are independent, and each has the same distribution as T_i, (where T_i is the time that the process first returns to the state i so we conclude from the theorems on renewal processes that

$$\lim_{t\to\infty}\frac{N_i(t)}{t}=\frac{1}{\mu_i}\text{ with probability 1,}$$

where $\mu_i = E[Z_n]$. Note that if state i is null recurrent, then $\mu_i = \infty$, and

$$\frac{1}{\mu_i}=0$$

and if i is positive recurrent, then $\mu_i < \infty$ and

$$\frac{1}{\mu_i}>0.$$

Also, from the elementary renewal theorem,

$$\lim_{t\to\infty}\frac{m_i(t)}{t}\to\frac{1}{\mu_i},$$

where $m_i(t) = E[N_i(t)]$.

By the theory of delayed renewal processes, we have the same results if the process begins in state j as long as the time of passage from state j to the first appearance of state i is finite with probability 1. From this, we get the following result.

Theorem 3.14

For a recurrent, irreducible, aperiodic Markov chain,

$$\lim_{n\to\infty} P^n(j,i) = \frac{1}{\mu_i}$$

for any states i and j.

Proof

The intuition of the proof is the observation that

$$\lim_{n\to\infty} P^n(j,i)$$

is the proportion of time that the process spends in state i having begun in state j. From delayed renewal theory, this is independent of state j under the hypotheses of the theorem, and renewal theory gives the result if the process begins in state i.

Recall that π is an equilibrium distribution for a Markov chain with transition matrix P if $\pi P = \pi$ with $\pi(i) \geq 0$ and $\sum \pi(i) = 1$. The condition $\pi P = \pi$ is

$$\pi(i) = \sum_j P(j,i)\pi(j).$$

Theorem 3.15

If $\{X_n\}$ is a positive recurrent, irreducible, aperiodic Markov chain, then $\pi = \pi(i)$, where

$$\pi(i) = \frac{1}{\mu(i)}$$

is an equilibrium distribution for $\{X_n\}$. Furthermore, this equilibrium state is unique.

Proof

Fix a state k. Since

$$\lim_{n\to\infty} P^n(k,j)$$

is the proportion of time that the process spends in state j having begun in state k, then

$$\sum_j \lim_{n\to\infty} P^n(k,j) = 1.$$

But

$$\lim_{n\to\infty} P^n(k,j) = \frac{1}{\mu(j)},$$

so

$$\sum_j \frac{1}{\mu(j)} = \sum_j \pi(j) = 1.$$

Now,

$$\frac{1}{\mu(i)} = \lim_{n\to\infty} P^n(k,i) = \lim_{n\to\infty} P^{n+1}(k,i),$$

and

$$\begin{aligned} P^{n+1}(k,i) &= P\{X_{n+1} = i | X_0 = k\} \\ &= \sum_j P\{X_{n+1} = i | X_n = j\} P\{X_n = j | X_0 = k\}. \end{aligned}$$

Also

$$P\{X_{n+1} = i | X_n = j\} = P(j,i) \quad \text{and} \quad P\{X_n = j | X_0 = k\} = P^n(k,j)$$

with

$$\lim_{n\to\infty} P^n(k,j) = \frac{1}{\mu(j)}.$$

Thus,

$$\frac{1}{\mu(i)} = \lim_{n\to\infty} P^n(k,i) = \lim_{n\to\infty} P^{n+1}(k,i) =$$

$$\lim_{n\to\infty} \sum_j P\{X_{n+1} = i \mid X_n = j\} P\{X_n = j \mid X_0 = k\} =$$

$$\lim_{n\to\infty} \sum_j P(j,i) P\{X_n = j \mid X_0 = k\} = \sum_j P(j,i) \lim_{n\to\infty} P\{X_n = j \mid X_0 = k\} =$$

$$\sum_j P(j,i) \lim_{n\to\infty} P^n(k,j) = \sum_j P(j,i) \frac{1}{\mu(j)}$$

where moving the limit inside the summation is justified because all quantities are nonnegative.

Thus,

$$\pi(i) = \sum_j P(j,i)\pi(j),\ \pi(i) \geq 0 \quad \text{and} \quad \sum_j \pi(j) = 1,$$

so π is an equilibrium distribution.

Uniqueness follows from our earlier observation that with the hypotheses of the theorem

$$\lim_{n\to\infty} P^n(i,j) = \frac{1}{\mu(j)}$$

independent of state i.

Difference Equations

Example

This example is somewhat long, but it has several interesting ideas including an application of the three previous theorems, and it illustrates determining the difference between null recurrence and positive recurrence. We first discuss solving a type of difference equation that will be needed.

A difference equation is the discrete-time analog of a differential equation.

The type of difference equation that we will need to solve is of the form

$$af(n+1)+bf(n)+cf(n-1)=0$$

where a, b, and c are constants. This type of equation is a second-order linear homogeneous equation with constant coefficients.

The procedure is to guess that the solutions are of the form $f(n) = m^n$ where the constants m are to be found.

In solving

$$af(n+1)+bf(n)+cf(n-1)=0,$$

setting $f(n) = m^n$ gives

$$am^{n+1}+bm^n+cm^{n-1}=0. \quad \text{So } m^{n-1}\left(am^2+bm+c\right)=0,$$

and we seek values of m for which $am^2 + bm + c = 0$.

Consider

$$f(n+1)-5f(n)+6f(n-1)=0.$$

Setting $f(n) = m^n$ gives

$$m^{n+1}-5m^n+6m^{n-1}=0. \quad \text{So } m^{n-1}\left(m^2-5m+6\right)=0$$

and $m = 2$ and 3. It is the case that

$$f(n)=2^n \quad \text{and} \quad f(n)=3^n$$

are linearly independent solutions, so that for any constants C_1 and C_2,

$$C_1 2^n + C_2 3^n$$

is a solution.

Like with ordinary differential equations, additional data, such as initial conditions, are needed to find C_1 and C_2. For example, if $f(0) = 0$ and $f(1) = -1$, then

$$C_1+C_2=0 \quad \text{and} \quad 2C_1+3C_2=-1, \quad \text{so } C_1=1 \text{ and } C_2=-1.$$

If there is only one value of m that satisfies

$$am^{n+1}+bm^n+cm^{n-1}=0,$$

then the solution is of the form

$$C_1 m^n + C_2 n m^n.$$

We apply these ideas to a random walk problem.

We begin with a Markov chain, with states {0, 1, 2, …}. We want to model a random walk with a partially reflecting barrier. (Having state 0 be reflecting or partially reflecting means there are no absorbing states.) Suppose that for some p, $0 < p < 1$, the process has the transition probabilities

$$p(j, j+1) = p, \quad j = 0, 1, 2, \ldots$$

$$p(j, j-1) = 1 - p, \quad j = 1, 2, \ldots$$

$$p(0,0) = 1 - p.$$

So the process moves to the right with probability p and to the left with probability $(1 - p)$.

Suppose that $\pi = (\pi(0), \pi(1), \pi(2), \ldots.)$ is a stationary probability distribution. If such a distribution exists, then the chain is positive recurrent.

The probability that the process is in state j at time $n > 0$ is denoted $P(X_n = j)$, and we have

$$P(X_n = j) = \sum_k P(X_{n-1} = k) p(k, j).$$

Now, if $(k, j) \neq (0, 0)$ we have

$$p(k,j) = \begin{cases} p & \text{if } k = j-1 \\ 1-p & \text{if } k = j+1 \\ 0 & \text{otherwise} \end{cases}$$

so

$$P(X_n = j) = P(X_{n-1} = j-1)p + P(X_{n-1} = j+1)(1-p).$$

In the stationary state, $P(X_k = j) = \pi(j)$, so we have

$$\pi(j) = p\pi(j-1) + (1-p)\pi(j+1), \; j \geq 1.$$

Thus, we solve

$$(1-p)\pi(j+1) - \pi(j) + p\pi(j-1) = 0.$$

Setting

$$\pi(j) = m^j$$

gives

$$(1-p)m^{j+1} - m^j + pm^{j-1} = m^{j-1}\left[(1-p)m^2 - m + p\right] = 0,$$

so

$$m = \frac{1 \pm \sqrt{1-4(1-p)p}}{2(1-p)} = \frac{1 \pm \sqrt{1-4p+4p^2}}{2(1-p)} = \frac{1 \pm (1-2p)}{2(1-p)}.$$

If $p \neq \frac{1}{2}$, the two roots are

$$m = \frac{1+(1-2p)}{2(1-p)} = 1 \quad \text{and} \quad m = \frac{1-(1-2p)}{2(1-p)} = \frac{p}{1-p},$$

and the solution is

$$\pi(j) = C_1 \cdot 1 + C_2 \left(\frac{p}{1-p}\right)^j. \tag{1}$$

Note that the series

$$\sum_{j=0}^{\infty} \left(\frac{p}{1-p}\right)^j$$

converges if

$$\left(\frac{p}{1-p}\right) < 1,$$

that is, if $p < \frac{1}{2}$ and diverges if $p > \frac{1}{2}$. (We are only considering the case $p \neq \frac{1}{2}$.)

Also, since

$$\sum_{j=1}^{\infty} \pi(j) = \sum_{j=1}^{\infty} \left[C_1 \cdot 1 + C_2 \left(\frac{p}{1-p}\right)^j\right] = 1,$$

we must have $C_1 = 0$.

So we have

$$\pi(j) = C_2 \left(\frac{p}{1-p} \right)^j.$$

To find C_2, we use

$$\sum_{j=0}^{\infty} \pi(j) = 1.$$

Now,

$$\sum_{j=0}^{\infty} \left(\frac{p}{1-p} \right)^j = \frac{1}{1 - \left(\frac{p}{1-p} \right)} = \frac{1-p}{1-2p},$$

so

$$C_2 = \frac{1-2p}{1-p}$$

and

$$\pi(j) = C_2 \left(\frac{p}{1-p} \right)^j = \left(\frac{1-2p}{1-p} \right) \left(\frac{p}{1-p} \right)^j.$$

Since we have found a distribution π for which $\pi P = \pi$ and

$$\sum_{j} \pi(j) = 1,$$

the process is positive recurrent if $p < \frac{1}{2}$.

If $p > \frac{1}{2}$, then

$$\left(\frac{p}{1-p} \right)^j \to \infty \text{ as } j \to \infty,$$

so the process is not positive recurrent.

We now apply Theorem 3.10 to find the values of p for which the process is transient. We set $z = 0$ in the definition of $\alpha(i)$.

$$\text{Since } \alpha(i) = P\{X_n = 0 \quad \text{for some } n \geq 0 | X_0 = i\},$$

$$\text{then } \alpha(i-1) = P\{X_n = 0 \quad \text{for some } n \geq 0 | X_0 = i-1\}, \quad \text{and}$$

$$\alpha(i+1) = P\{X_n = 0 \quad \text{for some } n \geq 0 | X_0 = i+1\}.$$

Now,

$$P\{X_n = 0 \quad \text{for some } n \geq 0 | X_0 = i\} = P\{X_n = 0 \quad \text{for some } n \geq 1 | X_0 = i\},$$

so

$$\begin{aligned} \alpha(i) &= P\{X_n = 0 \quad \text{for some } n \geq 0 | X_0 = i\} \\ &= P\{X_n = 0 \quad \text{for some } n \geq 1 | X_1 = i-1\} P\{X_1 = i-1 | X_0 = i\} \\ &\quad + P\{X_n = 0 \quad \text{for some } n \geq 1 | X_1 = i+1\} P\{X_1 = i+1 | X_0 = i\} \\ &= \alpha(i-1)(1-p) + \alpha(i+1)p. \end{aligned}$$

Note this is different from

$$\pi(j) = p\pi(j-1) + (1-p)\pi(j+1), \; j \geq 1$$

that we had for π being the stationary state.

We solve $\alpha(i) = \alpha(i-1)(1-p) + \alpha(i+1)p$ or

$$p\alpha(i+1) - \alpha(i) + (1-p)\alpha(i-1) = 0.$$

Setting $\alpha(i) = m^i$, we have

$$pm^{i+1} - m^i + (1-p)m^{i-1} = m^{i-1}\left(pm^2 - m + (1-p)\right) = 0$$

so

$$m = \frac{1 \pm \sqrt{1 - 4p(1-p)}}{2p} = \frac{1 \pm (1-2p)}{2p}$$

and

$$m = \frac{1-p}{p} \quad \text{and} \quad m = 1.$$

Again, contrast this with the case where we computed π and found

$$m = \frac{p}{1-p} \quad \text{and} \quad m = 1.$$

Thus,

$$\alpha(i) = C_1 + C_2\left(\frac{1-p}{p}\right)^i \quad \text{if } p \neq \frac{1}{2}$$

and

$$\alpha(i) = C_1 + C_2 i \quad \text{if } p = \frac{1}{2}.$$

We first consider the $p \neq \frac{1}{2}$ case. Since α(0) = 1, we have

$$1 = C_1 + C_2 \quad \text{so } C_1 = 1 - C_2 \text{ and}$$

$$\alpha(i) = (1 - C_2) + C_2\left(\frac{1-p}{p}\right)^i.$$

$$\text{If } p > \frac{1}{2} \quad \text{and} \quad C_2 = 1, \text{ then}$$

$$\alpha(i) = \left(\frac{1-p}{p}\right)^i,$$

and we potentially have a solution. We must still check

$$\text{for } i \neq j, \quad \alpha(i) = \sum_{k \epsilon S} p(i,k)\alpha(k).$$

Now, $p(i,k) = 0$ unless $k = i + 1$ or $k = i - 1$, so for this to be a solution, we must have

$$\alpha(i) = p(i,i+1)\alpha(i+1) + p(i,i-1)\alpha(i-1).$$

Now,

$$\begin{aligned}
&p(i,i+1)\alpha(i+1) + p(i,i-1)\alpha(i-1) \\
&= p\left(\frac{1-p}{p}\right)^{i+1} + (1-p)\left(\frac{1-p}{p}\right)^{i-1} \\
&= \left(\frac{1-p}{p}\right)^{i-1}\left[p\left(\frac{1-p}{p}\right)^2 + (1-p)\right] = \left(\frac{1-p}{p}\right)^{i-1}(1-p)\left[p\frac{1-p}{p^2} + 1\right] \\
&= \left(\frac{1-p}{p}\right)^{i-1}(1-p)\left[\frac{1-p}{p} + 1\right] = \left(\frac{1-p}{p}\right)^{i-1}(1-p)\frac{1}{p} = \left(\frac{1-p}{p}\right)^i = \alpha(i).
\end{aligned}$$

Thus, the chain is transient if $p > \frac{1}{2}$.

It remains to check the case $p = \frac{1}{2}$.
We return to the equation

$$\pi(x) = C_1 + C_2 x.$$

We had determined

$$\pi(0) = \frac{1-p}{p}\pi(1) \quad \text{and} \quad \frac{1-p}{p}\pi(1) = \pi(1) \quad \text{if } p = \frac{1}{2},$$

$$\text{so } \pi(0) = \pi(1) \quad \text{if } p = \frac{1}{2}.$$

Now,

$$\pi(0) = C_1 \quad \text{and} \quad \pi(1) = C_1 + C_2, \quad \text{so if } \pi(0) = \pi(1), \quad \text{then } C_2 = 0.$$

Thus, $\pi(i) = C_1$ satisfies $\pi P = \pi$, but π cannot be a probability distribution because

$$\sum_{j=1}^{\infty} C_1 = \begin{cases} 0 & \text{if } C_1 = 0 \\ \infty & \text{if } C_1 \neq 0' \end{cases}$$

so if $p = \frac{1}{2}$, the chain is null recurrent.

We have very little to say about periodic infinite Markov processes. The one result that we mention without proof is that if $\{X_n\}$ is an irreducible Markov chain of period d, then

$$\lim_{n\to\infty} P^{nd}(i,i) = \frac{d}{\mu(i)}.$$

Branching Processes

Branching processes can be understood via the following problem, known as the Galton–Watson process.

We begin with a single entity that can reproduce and keep track of the offspring. Let p_k be the probability that one individual will produce exactly k

offspring during its lifetime. We assume that $p_0 > 0$ so that extinction is possible and that $p_0 + p_1 < 1$ so that population growth is possible. We further assume that these probabilities are independent and identically distributed for all individuals. We want to determine the probability the family will become extinct.

The analysis will use generating functions.

In Chapter 1, we presented the material for probability generating functions that will be used in this section. We recall some of the pertinent properties now.

For a branching process, the random variables are nonnegative integer valued, and in this setting, it is convenient to use the probability generating function defined by

$$G(z) = E\left[z^X\right] = \sum_{n=0}^{\infty} P(X = n) z^n, \quad 0 \le z \le 1.$$

Then

$$G'(1) = E[X], \quad G''(1) = E[X(X-1)],$$

so

$$\begin{aligned} G''(1) + G'(1) - \left(G'(1)\right)^2 &= E[X(X-1)] + E[X] - \left(E[X]\right)^2 \\ &= E\left[X^2\right] - E[X] + E[X] - \left(E[X]\right)^2 = \mathrm{Var}(X). \end{aligned}$$

We will use the fact that if X and Y are independent random variables with generating functions $G_X(z)$ and $G_Y(z)$, then the generating function of $X + Y$ is $G_X(z)G_Y(z)$.

With the Galton–Watson process,

$$G(z) = \sum_{k=0}^{\infty} p_k z^k, \quad 0 \le z \le 1$$

where the p_k's are as described earlier.

We now describe the Galton–Watson process as a Markov chain. We refine our description of the process to assume that in a given generation, all members of that generation have all their offspring simultaneously and die immediately after giving birth. We let X_n be the size of the nth generation. We have assumed that the process begins with exactly one member, so $X_0 = 1$.

The transition probabilities are not easily formulated, and, in fact, we will not need them to answer the questions we have posed.

Let $G_n(z)$ be the probability generating function for X_n; that is,

$$G_n(z) = E\left[z^{X_n}\right] = \sum_{k=0}^{\infty} P\{X_n = k\} z^k.$$

Then

$$G_1(z) = E\left[z^{X_1}\right] = \sum_{k=0}^{\infty} z^k P(X_1 = k) = \sum_{k=0}^{\infty} p_k z^k = G(z).$$

We will use that $G(z)$ is continuous and increasing for $0 \le z \le 1$.

Theorem 3.16

Let $G_k(z)$ be as that given earlier. Then $G_n(z) = G\left(G_{n-1}(z)\right)$.

Proof

A key idea of the proof is to compare X_{n-1} given $X_0 = j$ with X_{n-1} given $X_0 = 1$. We have that X_{n-1} given $X_0 = j$ is the sum of j independent random variables, each of whose distribution is X_{n-1} given $X_0 = 1$. So

$$\sum_{k=0}^{\infty} P\{X_{n-1} = k | X_0 = j\} z^k = j \sum_{k=0}^{\infty} P\{X_{n-1} = k | X_0 = 1\} = \left[G_{n-1}(z)\right]^j$$

since if X and Y are independent random variables with generating functions $G_X(z)$ and $G_Y(z)$, then the generating function of $X + Y$ is $G_X(z)G_Y(z)$.

Thus,

$$G_n(z) = E\left[z^{X_n}\right] = \sum_{k=0}^{\infty} P(X_n = k) z^k$$

$$= \sum_{k=0}^{\infty} \left[\sum_{j=0}^{\infty} P\{X_n = k | X_1 = j\} P(X_1 = j) \right] z^k$$

$$= \sum_{j=0}^{\infty} P(X_1 = j) \left[\sum_{j=0}^{\infty} P\{X_{n-1} = k | X_0 = j\} z^k \right]$$

$$= \sum_{j=0}^{\infty} p_j \left[G_{n-1}(z)\right]^j = G\left(G_{n-1}(z)\right).$$

Corollary

For n, a positive integer, $G_n(z) = G\left(G\left(\cdots G(z)\right)\right)$.

Thus, we have

$$G_n(z) = G\left(G\left(\cdots G(z)\right)\right) = E\left[z^{X_n}\right].$$

This means $P\{X_n = 0 | X_0 = 1\} = G_n(0)$, so that $G_n(0)$ is the probability that extinction has occurred by the nth generation. Note that $\{G_n(0)\}$ is an increasing sequence that is bounded above and the probability of extinction is

$$\lim_{n \to \infty} G_n(0).$$

Let E be the event the population becomes extinct and $\varepsilon = P(E)$, the probability of extinction. We have $\varepsilon = G(\varepsilon)$ since

$$\varepsilon = P\{E | X_0 = 1\} = \sum_{k=0}^{\infty} P\{X_1 = k | X_0 = 1\} P\{E | X_1 = k\}$$

$$= \sum_{k=0}^{\infty} p_k \varepsilon^k = G(\varepsilon).$$

The equality $P\{E | X_1 = k\} = \varepsilon^k$ follows because if at a certain time there are k individuals, we can consider this as a process that has k *branches*, and for the *tree* to become extinct, all the branches must become extinct, and the branches are independent.

Thus, the extinction probability ε satisfies the equation $z = G(z)$, as does 1.

A major result is Theorem 3.17.

Theorem 3.17

The extinction probability, ε, is the smallest positive root of $z = G(z)$.

Proof

We have shown that $\varepsilon = G(\varepsilon)$. Suppose that r is the smallest positive root of $z = G(z)$. We show that $\varepsilon \le r$ by showing that for every nonnegative integer n,

$$P\left(X_n = 0\right) \le r.$$

This will mean

$$\varepsilon = \lim_{n\to\infty} P\left\{X_n = 0 | X_0 = 1\right\} \le r$$

and thus $\varepsilon = r$.

We do induction on n.

Since $P\{X_0 = 0 | X_0 = 1\} = 0 \le r$, the result holds when $n = 0$.

Assume the result holds when $n = k$. Now, since G is increasing,

$$\left\{X_{n+1} = 0 | X_0 = 1\right\} = G_{n+1}\left(0\right) = G\left(G_n\left(0\right)\right) \le G\left(r\right) = r$$

and thus the result is true for all nonnegative integers.

One would expect that the probability of extinction would be related to the mean number of offspring of an individual that is

$$\mu = \sum_{k=0}^{\infty} kp_k = E\left[X_1\right].$$

Note that

$$G'\left(z\right) = \sum_{k=0}^{\infty} p_k k z^{k-1}, \quad \text{so that } \mu = G'\left(1\right).$$

If $X_n = k$, then X_{n+1} will have the same distribution as the sum of k random variables, each of which has the same distribution as X_0, so $E\left[X_{n+1} \middle| X_n = k\right] = k\mu$. Also,

$$E[X_n] = \sum_{k=0}^{\infty} E\left[X_n \middle| X_{n-1} = k\right] P(X_{n-1} = k) = \sum_{k=0}^{\infty} k\mu P(X_{n-1} = k)$$

$$= \mu \sum_{k=0}^{\infty} kP(X_{n-1} = k) = \mu E[X_{n-1}]$$

so that $E[X_n] = \mu^n E[X_0] = \mu^n$.

From this, we get

$$\lim_{n\to\infty} E[X_n] = \begin{cases} 0 & \text{if } \mu < 1 \\ 1 & \text{if } \mu = 1. \\ \infty & \text{if } \mu > 1 \end{cases}$$

Theorem 3.18

If $\mu \le 1$, then $\lim_{n\to\infty} P(X_n = 0) = 1$.

Proof

We have

$$P(X_n = 0) = G_n(0) = G\left(G\left(\cdots G(0)\right)\right), \quad G(0) = p_0 > 0, G(1) = 1, G'(1) = \mu.$$

Also

$$G'(z) = \sum_{k=0}^{\infty} kp_k z^{k-1} = \sum_{k=1}^{\infty} kp_k z^{k-1} \ge 0,$$

so $G(z)$ is increasing and

$$G''(z) = \sum_{k=1}^{\infty} k(k-1)p_k z^{k-2} = \sum_{k=2}^{\infty} k(k-1)p_k z^{k-2} \ge 0,$$

so $G(z)$ is concave up.

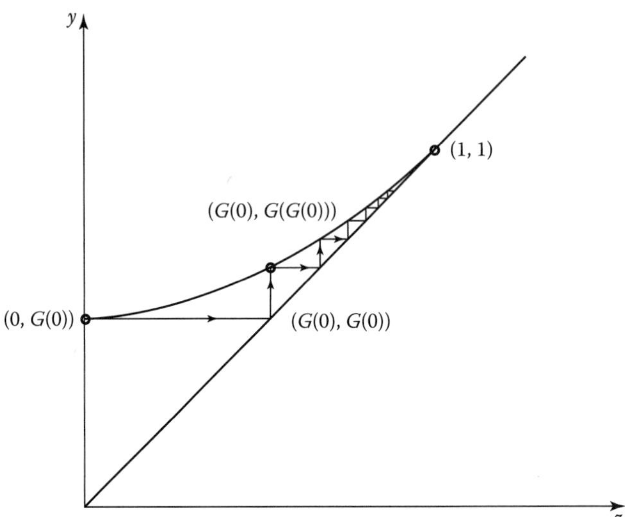

FIGURE 3.2
Graphs of $y = G(z)$ and $y = z$ for $\mu \le 1$.

Suppose that $\mu \le 1$. We compare the graphs of $y = z$ and $y = G(z)$. Now, $G'(z) \le G'(1) = \mu \le 1$ for $z \le 1$, so $G(z)$ is increasing less rapidly than z. This together with $G(1) = 1$ means the graph of $G(z)$ is above the graph of z for $z < 1$. See Figure 3.2.

We discuss how to plot the iterates $G_n(0) = G(G(\cdots G(0)))$. Starting at $z = 0$, draw a vertical line to the graph of $G(z)$. The y-coordinate of this point is $G(0)$. To compute $G(G(0))$, we set $z = G(0)$ by drawing a horizontal line from $(0, G(0))$ to $(G(0), G(0))$, that is, to the graph of $y = z$. We then draw a vertical line to the graph of $G(z)$. The y-coordinate of this point is $G(G(0))$. As n increases, $G_n(0)$ converges to where the graph of $y = z$ intersects with the graph of $y = G(z)$, which in this case is at (1,1). Thus,

$$\lim_{n \to \infty} P(X_n = 0) = \lim_{n \to \infty} G_n(0) = 1.$$

Theorem 3.19

If $\mu > 1$, then $\lim_{n\to\infty} P(X_n = 0) < 1$.

Proof

We again have

$$P(X_n = 0) = G_n(0) = G(G(\cdots G(0))), \quad G(0) = p_0 > 0, G(1) = 1, G'(1) = \mu$$

and that $G(z)$ is increasing and concave up. However, since $G'(1) = \mu > 1$, the graph of $G(z)$ is increasing more rapidly than the graph of z at $z = 1$.

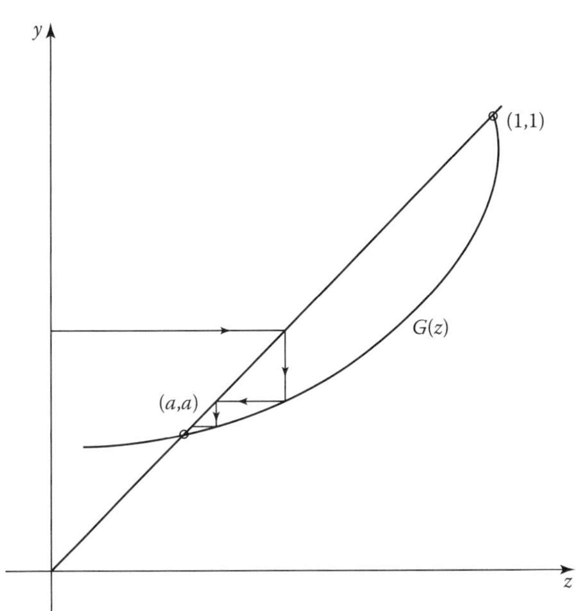

FIGURE 3.3
Graphs of $y = G(z)$ and $y = z$ for $\mu > 1$.

This means there is an interval $(a,1)$ where the graph of $G(z)$ is below the graph of z. The number a is the one value between 0 and 1 where $G(a) = a$. See Figure 3.3.

Now, when we iterate $G(z)$ repeatedly, the process converges to (a,a), so $\lim_{n\to\infty} P(X_n = 0) = a < 1$.

Example

With the aid of a CAS, we can find the probability of extinction.

1. Let $p_0 = 1/8,\ \ p_1 = 1/4,\ \ p_2 = 1/2,\ \ p_3 = 1/8$.
   We can find the probability of extinction either by solving

$$x = \frac{1}{8} + \frac{1}{4}x + \frac{1}{2}x^2 + \frac{1}{8}x^3$$

   or by determining where the graphs of x and $\frac{1}{8} + \frac{1}{4}x + \frac{1}{2}x^2 + \frac{1}{8}x^3$ intersect. (The answer is $a \approx .193$.) In this example,

$$\mu = 0 \cdot \frac{1}{8} + 1 \cdot \frac{1}{4} + 2 \cdot \frac{1}{2} + 3 \cdot \frac{1}{8} = \frac{13}{8} > 1,$$

   so the probability of extinction is less than 1.

2. Let $p_0 = 1/2,\ \ p_1 = 1/4,\ \ p_2 = 1/4$. In this example,

$$\mu = 0\cdot\frac{1}{2} + 1\cdot\frac{1}{4} + 2\cdot\frac{1}{4} = \frac{3}{4} < 1,$$

so the probability of extinction is 1. In fact, the smallest positive solution to

$$x = \frac{1}{2} + \frac{1}{4}x + \frac{1}{4}x^2$$

is 1.

Random Walk in $\mathbb{Z}^d$

Suppose that a particle starts at the origin in $\mathbb{Z}^d$, and at each unit of time, it moves to one of its $2d$ nearest neighbors with equal probability; that is, with probability $1/2d$. We investigate whether the particle returns to the origin infinitely often with probability 1. The answer depends on the value of d.

Theorem 3.20

Suppose the probability of a particle moving in any direction on the lattice $\mathbb{Z}^d$ is equally likely. If $d = 1$ or 2, a random walk in $\mathbb{Z}^d$ is recurrent, and if $d > 2$, the random walk is transitive.

Proof

We will use Stirling's approximation to estimate factorials in the proof. It says

$$n! \sim \sqrt{2\pi n}\, e^{-n}\, n^n$$

by which we mean

$$\lim_{n\to\infty} \frac{n!}{\sqrt{2\pi n}\, e^{-n}\, n^n} = 1.$$

Proof for $d = 1$:

The process can return to the origin only after an even number of steps.

If $2n$ steps have been taken and the process is at the origin, then there must have been n steps taken to the left and n to the right. This can be done in

$$\binom{2n}{n} = \frac{(2n)!}{n!n!}$$

ways. If the probability of moving to the left is ½ (so the probability of moving to the right is also ½), then the probability of any given path of length $2n$ is

$$\left(\frac{1}{2}\right)^{2n}.$$

Thus to be at the origin after $2n$ jumps, we have $\binom{2n}{n}$ possible paths, each with probability $\left(\frac{1}{2}\right)^{2n}$, so the probability the process is at 0 after $2n$ jumps is

$$\binom{2n}{n}\left(\frac{1}{2}\right)^{2n} = \frac{(2n)!}{n!n!}\left(\frac{1}{2}\right)^{2n} \sim \frac{\sqrt{2\pi(2n)}\ e^{-2n}(2n)^{2n}}{(\sqrt{2\pi n}e^{-n}\ n^n)^2}\left(\frac{1}{2}\right)^{2n} = \frac{1}{\sqrt{\pi n}}.$$

Since

$$\sum_{n=1}^{\infty}\frac{1}{\sqrt{\pi n}}$$

diverges, the state is recurrent.

Proof for $d = 2$:

In two dimensions, in order for the process to return to the origin, it must have taken the same number of steps up as down and the same number of steps to the left as to the right. As before, this is possible only if there are an even number of steps, so consider the case where there are $2n$ steps. Now, there are 4 choices for the particle in making a jump, so there are 4^{2n} paths of length $2n$, and each occurs with probability $(1/4)^{2n}$. In order for the process to return to the origin after $2n$ steps, there must be k steps to the left and k steps to the right $0 \le k \le n$ and $(n - k)$ steps up and $(n - k)$ steps down. The number of such paths for a fixed, admissible k is

$$\binom{2n}{k,k,n-k,n-k} = \frac{(2n)!}{k!k!(n-k)!(n-k)!},$$

so the total number of paths that return to the origin after $2n$ steps is

$$\sum_{k=0}^{n}\frac{(2n)!}{k!k!(n-k)!(n-k)!}=\sum_{k=0}^{n}\frac{(2n)!}{n!n!}\cdot\frac{n!n!}{k!k!(n-k)!(n-k)!}$$

$$=\frac{(2n)!}{n!n!}\sum_{k=0}^{n}\frac{n!n!}{k!k!(n-k)!(n-k)!}=\frac{(2n)!}{n!n!}\sum_{k=0}^{n}\left[\frac{n!}{k!(n-k)!}\right]^2$$

$$=\frac{(2n)!}{n!n!}\sum_{k=0}^{n}\binom{n}{k}^2=\binom{2n}{n}\sum_{k=0}^{n}\binom{n}{k}^2.$$

We give a derivation of the formula

$$\binom{2n}{n}=\sum_{k=0}^{n}\binom{n}{k}^2.$$

Suppose we have m men and n women and we want to form a committee of r members. The number of possible committees could be computed in two ways:

1. $\binom{m+n}{r}$
2. The number of committees with k men and $r-k$ women is $\binom{m}{k}\binom{n}{r-k}$.

The sum over the admissible values of k to get the number of committees is

$$\sum_{k=0}^{r}\binom{m}{k}\binom{n}{r-k}.$$

Thus,

$$\binom{m+n}{r}=\sum_{k=0}^{r}\binom{m}{k}\binom{n}{r-k}.$$

Letting $m=r=n$ yields

$$\binom{m+n}{r}=\binom{2n}{n}\quad\text{and}\quad\sum_{k=0}^{r}\binom{m}{k}\binom{n}{r-k}=\sum_{k=0}^{n}\binom{n}{k}\binom{n}{n-k},$$

so

$$\binom{2n}{n}=\sum_{k=0}^{n}\binom{n}{k}\binom{n}{n-k}.$$

Now,

$$\binom{n}{k}\binom{n}{n-k}=\left[\frac{n!}{k!(n-k)!}\right]\left[\frac{n!}{(n-k)!k!}\right]=\left[\frac{n!}{k!(n-k)!}\right]^2=\binom{n}{k}^2,$$

and we have

$$\binom{2n}{n}=\sum_{k=0}^{n}\binom{n}{k}^2.$$

Thus, the probability of returning to (0,0) after $2n$ steps is

$$\left(\frac{1}{4}\right)^{2n}\left[\binom{2n}{n}\sum_{k=0}^{n}\binom{n}{k}^2\right]=\left(\frac{1}{4}\right)^{2n}\binom{2n}{n}^2=\left(\frac{1}{2}\right)^{4n}\binom{2n}{n}^2=\left[\left(\frac{1}{2}\right)^{2n}\binom{2n}{n}\right]^2.$$

In the one dimensional case, we showed

$$\binom{2n}{n}\left(\frac{1}{2}\right)^{2n}\sim\frac{1}{\sqrt{\pi n}}$$

so

$$\left[\left(\frac{1}{2}\right)^{2n}\binom{2n}{n}\right]^2\sim\frac{1}{\pi n}$$

and

$$\sum_{n=1}^{\infty}\frac{1}{\pi n}$$

diverges so the system is recurrent.

Proof for $d = 3$:

The proofs for $d = 1$ and $d = 2$ provide guidance for what to expect in higher dimensions. We give an intuitive argument for the general case $d \geq 3$ and then a more rigorous argument that validates the method for $d = 3$.

If we are in d dimensions and make $2n$ steps, we would expect (according to the law of large numbers) that approximately $2n/d$ of those jumps are made parallel to each axis. Name the axes $x_1, x_2, \ldots, x_d$. Partition the jumps according to the dimension in which they took place. In order for the process to return to the origin, it must be the case that in each dimension, the number of jumps in one direction is the same as the number of jumps in the opposite direction. Fix a particular axis and consider only the jumps made parallel to that axis. This harkens back to the $d = 1$ case where we showed the probability of returning to the origin after $2n$ steps is asymptotically $C/\sqrt{n}$. What happens in a particular dimension is independent of what happens in other dimensions, and in order for the process to return to the origin, it must be at the origin of each dimension. Thus, the probability of being at the origin after $2n$ steps is asymptotically

$$\frac{C}{\sqrt{n}} \cdot \frac{C}{\sqrt{n}} \cdot \cdots \cdot \frac{C}{\sqrt{n}} = \frac{K}{n^{d/2}}.$$

The process will be recurrent if

$$\sum_{n=1}^{\infty} \frac{1}{n^{d/2}}$$

diverges, and this occurs if $d \leq 2$ and is transient if the series converges, which occurs if $d > 2$.

For $d = 3$, a more specific argument is that since the process has 6 choices where to move at each step, every path of length $2n$ has a probability of

$$\left(\frac{1}{6}\right)^{2n}$$

occurring.

The number of paths with k steps to the left, k steps to the right, j steps up, j steps down, $(n - j - k)$ steps forward, and $(n - j - k)$ steps backward is

$$\binom{2n}{k, k, j, j, n-j-k, n-j-k} = \frac{(2n)!}{k!k!j!j!(n-j-k)!(n-j-k)!},$$

so the number of admissible paths is

$$\sum_{j,k\geq 0,0\leq j+k\leq n} \frac{(2n)!}{k!k!j!j!(n-j-k)!(n-j-k)!}$$

$$= \sum_{j,k\geq 0,0\leq j+k\leq n} \frac{(2n)!}{n!n!} \cdot \frac{n!n!}{k!k!j!j!(n-j-k)!(n-j-k)!}$$

$$= \binom{2n}{n} \sum_{j,k\geq 0,0\leq j+k\leq n} \left[\frac{n!}{k!j!(n-j-k)!}\right]^2.$$

The maximum value of

$$\frac{n!}{k!j!(n-j-k)!}$$

is attained when j, k, and $(n-j-k)$ are as close as possible to being equal. Thus,

$$\frac{n!}{k!j!(n-j-k)!} \leq \frac{n!}{\left(\left[\frac{n}{3}\right]!\right)^3}$$

where [ ] is the greatest integer function. So

$$\left[\frac{n!}{k!j!(n-j-k)!}\right]^2 \leq \frac{n!}{\left(\left[\frac{n}{3}\right]!\right)^3} \cdot \frac{n!}{k!j!(n-j-k)!},$$

and thus,

$$\binom{2n}{n} \sum_{j,k\geq 0,0\leq j+k\leq n} \left[\frac{n!}{k!j!(n-j-k)!}\right]^2$$

$$\leq \binom{2n}{n} \frac{n!}{\left(\left[\frac{n}{3}\right]!\right)^3} \sum_{j,k\geq 0,0\leq j+k\leq n} \frac{n!}{k!j!(n-j-k)!}.$$

To simplify further, note that

$$\sum_{j,k\geq 0, 0\leq j+k\leq n} \frac{n!}{k!j!(n-j-k)!}$$

is the number of ways that n objects may be distributed into 3 containers. Since there are 3^n ways of doing this,

$$\sum_{j,k\geq 0, 0\leq j+k\leq n} \frac{1}{3^n} \frac{n!}{k!j!(n-j-k)!} = 1,$$

so

$$\binom{2n}{n} \frac{n!}{\left(\left[\frac{n}{3}\right]!\right)^3} \sum_{j,k\ 0\leq j+k\leq n} \frac{n!}{k!j!(n-j-k)!}$$

$$= \binom{2n}{n} \frac{n!}{\left(\left[\frac{n}{3}\right]!\right)^3} 3^n \sum_{j,k\ 0\leq j+k\leq n} \frac{1}{3^n} \cdot \frac{n!}{k!j!(n-j-k)!}$$

$$= \binom{2n}{n} \frac{n!}{\left(\left[\frac{n}{3}\right]!\right)^3} 3^n.$$

Thus, the probability of returning to the origin after $2n$ jumps is approximately

$$\left(\frac{1}{6}\right)^{2n} \binom{2n}{n} \sum_{j,k\geq 0,\ 0\leq j+k\leq n} \left[\frac{n!}{k!j!(n-j-k)!}\right]^2 \leq \left(\frac{1}{6}\right)^{2n} \binom{2n}{n} \frac{n!}{\left(\left[\frac{n}{3}\right]!\right)^3} 3^n$$

$$= \left(\frac{1}{2}\right)^{2n} \binom{2n}{n} \frac{n!}{\left(\left[\frac{n}{3}\right]!\right)^3}.$$

Now,

$$\left(\frac{1}{2}\right)^{2n}\binom{2n}{n} \sim \frac{1}{\sqrt{\pi n}},$$

and by Stirling's formula and the fact that $\sqrt{\left[\frac{n}{3}\right]} \approx \sqrt{\frac{n}{3}}$ for n large, we have

$$\frac{n!}{\left(\left[\frac{n}{3}\right]!\right)^3} \sim \frac{\sqrt{2\pi n}\, n^n e^{-n}}{\left[\sqrt{2\pi\frac{n}{3}}\ n^n e^{-n}\right]^3} \approx \frac{3^{2n}3\sqrt{3}}{2\pi n}.$$

Thus, the probability of returning to the origin after $2n$ jumps is less than or equal to

$$\frac{C}{n^{3/2}}.$$

Since the series

$$\sum_{n=1}^{\infty}\frac{C}{n^{3/2}}$$

converges, the process is transient.

Example

We show that a random walk in one dimension is transient if $p \neq 1/2$. The intuition is that a particle will drift in the direction for which $p > 1/2$.

We repeat the argument for the one-dimensional case given in the theorem, except now the probability of a path of length $2n$ making n jumps in one direction and n jumps in the opposite direction is

$$p^n(1-p)^n$$

so that the probability of a path returning to the origin after $2n$ jumps is

$$\binom{2n}{n}p^n(1-p)^n.$$

We showed in the case $d = 1$

$$\frac{(2n)!}{n!n!} \sim \frac{\sqrt{2\pi(2n)}\ e^{-2n}(2n)^{2n}}{(\sqrt{2\pi n}e^{-n}\, n^n)^2} = \frac{2^{2n}}{\sqrt{n\pi}}$$

so the probability of a path returning to the origin after $2n$ jumps is

$$\frac{\left[4p(1-p)\right]^n}{\sqrt{n\pi}}.$$

But if $p \neq 1/2$, then $4p(1-p) < 1$, so the series

$$\sum_{n=1}^{\infty} \frac{\left[4p(1-p)\right]^n}{\sqrt{n\pi}}$$

converges and the chain is transient.

Exercises

3.1 Suppose an event occurs at times $t = .1, 3.6, 5.2$, and 9.1. Find

$$X_n,\ S_n,\ N(t), \text{ and } S_{N(t)}.$$

3.2 Suppose an event occurs at times $t = 4.4, 8.7, 9.1$, and 12.5. Find

$$X_n,\ S_n,\ N(t), \text{ and } S_{N(t)}.$$

3.3 If each X_n has the cumulative density function

$$F(x) = 1 - e^{-x}, \quad x > 0,$$

find F_2 and F_3 where $F_1 = F$ and

$$F_{k+1}(x) = \sum_{y=0}^{x} F_k(x-y) f(y).$$

3.4 If each X_n has the cumulative density function

$$F(x) = 1, \quad 0 < x < 1,$$

find F_2 and F_3.

3.5 (a) Show

$$x \in \limsup E_n$$

if and only if $x \in E_n$ for infinitely many E_n.

(b) Show

$$x \in \liminf E_n$$

if and only if $x \in E_n$ for all but finitely many E_n.

(c) Show

$$\liminf E_n \subset \limsup E_n.$$

(d) Give an example where

$$\liminf E_n \neq \limsup E_n.$$

(e) Show

$$\left(\limsup E_n\right)^c = \liminf\left(E_n{}^c\right).$$

3.6 Suppose

$$P\left(X_n = k\right) = e^{-\lambda}\frac{\lambda^k}{k!}, \quad \lambda > 0,\ k = 0, 1, 2, \ldots.$$

$$\text{Compute } P\left(N\left(t\right) = n\right).$$

3.7 Suppose that the expected lifetime of an electrical component is 1000 h. Compute

$$\lim_{t\to\infty}\frac{N(t)}{t}.$$

3.8 Let X_n be the number of rolls between 6's on a die.

(a) Find $P(X_n = k)$.

(b) Find $P(S_n = m)$.

(c) Find $P(N(t) = j)$.

3.9 Let $\{X_n\}$ be a sequence of independent identically distributed Bernoulli random variables with parameter p. We do a series of experiments with X_n being the outcome of the nth trial.

(a) Let $\{Y_n\}$ be the sequence of random variables where Y_n is the number of successes in the first n trials. What is the distribution of Y_n? What is $P(Y_n = k)$?

(b) Let $\{Z_n\}$ be the sequence of random variables where Z_n is the number of trials to go from the $(n-1)$st success to the nth success. What is the distribution of Z_n? What is crucial about this distribution as far as a process being a renewal process? What is $P(Z_n = k)$?

(c) Let $\{W_n\}$ be the sequence of random variables where W_n is the number of trials required to achieve n successes. What is the distribution of W_n? What is $P(Z_n = k)$?

3.10 This exercise is a variation on the random walk in $\mathbb{Z}$. The state space is $\mathbb{Z}$, but the transition probabilities have the effect of making 0 the preferred state. Let

$$p(0,1) = p(0,-1) = \frac{1}{2};$$

$$\text{if } n > 0,\quad p(n,n+1) = \frac{1}{3},0,\quad p(n,n-1) = \frac{2}{3};$$

$$\text{if } n < 0,\quad p(n,n+1) = \frac{2}{3},0,\quad p(n,n-1) = \frac{1}{3}.$$

(a) Without doing any calculations, do you believe the Markov chain is transient or recurrent? If your answer is recurrent, do you believe it is positive recurrent or null recurrent?

(b) Do the calculations confirm or reject your conjecture? If it is positive recurrent, find the equilibrium state.

3.11 This problem is a random walk with a bungee cord. The state space is $\{0, 1, 2, \ldots\}$, and the process usually moves to the right, but every so often, it returns to 0. The long-range behavior depends on how often the process returns to 0. In the following problems, determine if the Markov chain is transient, null recurrent, or positive recurrent. If it is positive recurrent, give the equilibrium state. Compare the likelihood of returning to 0 from a large n in the three cases.

(a) $p(n,n+1)=\frac{n^2}{n^2+1},\quad p(n,0)=\frac{1}{n^2+1}$

(b) $p(n,n+1)=\frac{n}{n+1},\quad p(n,0)=\frac{1}{n+1}$

$$p(n,n+1)=\frac{1}{n+1},\quad p(n,0)=\frac{n}{n+1}.$$

3.12 In a branching process with $\mu < 1$, find the expected number of family members of all generations before the family becomes extinct.

3.13 In a branching process with

$$p_0=\frac{1}{2},\quad p_1=\frac{1}{4},\quad p_2=\frac{1}{4},$$

find the probability

(a) The family dies out in the second generation; that is, $X_2 = 0$ but $X_1 \neq 0$

(b) The family dies out in the third generation; that is, $X_3 = 0$ but $X_2 \neq 0$

3.14 Give an expression for the expected time of extinction for a branching process in terms of G_n.

3.15 For the following branching processes, find μ and the probability of extinction:

(a) $p_0=\frac{1}{2},\ p_1=\frac{3}{8},\ p_2=\frac{1}{8}.$

(b) $p_0=\frac{1}{8},\ p_1=\frac{3}{8},\ p_2=\frac{1}{2}.$

(c) $p_0=\frac{1}{2},\ p_1=\frac{3}{8},\ p_{16}=\frac{1}{8}.$

3.16 We have a branching process with

$$p_i=(1-q)q^i,\quad 0<q<1.$$

(a) What is μ?

(b) How does the probability of extinction vary as q varies?

(c) Find the probability of extinction. It may be helpful to note that the series $p_0 + p_1x + p_2x^2 + \cdots$ can be written as a geometric series.

4

Exponential Distribution and Poisson Process

Continuous Random Variables

The purpose of this chapter is to provide the background necessary to study continuous-time Markov chains. Our starting point is a discussion of continuous random variables. Up to this time, the random variables that we have studied have had a discrete probability density function (p.d.f.). We now take up the case where the p.d.f. is continuous.

Definition

We say that the random variable X is continuous if there is a function $f(x) \geq 0$ such that for every set E of real numbers

$$P(X \in E) = \int_E f(x)\,dx.$$

The function $f(x)$ is the p.d.f. for X.

To be a p.d.f., $f(x)$ must satisfy

$$\int_E f(x)\,dx \geq 0 \quad \text{for every set } E$$

and

$$\int_{-\infty}^{\infty} f(x)\,dx = 1.$$

The intuition of the integral is $P(X \in [x, x + \Delta x]) \approx f(x)\,\Delta x$.

NOTE: In more advanced settings one correctly makes the restriction that the set is "measurable" but this will not affect our development.

We have the properties

$$P(a \le X \le b) = \int_a^b f(x)dx$$

and

$$P(X = a) = \int_a^a f(x)dx = 0, \quad \text{so } P(a \le X \le b) = P(a < X < b).$$

Analogous with the discrete case, the expected value of X is given by

$$E[X] = \int_{-\infty}^{\infty} xf(x)dx,$$

provided

$$\int_{-\infty}^{\infty} |x|\, f(x)dx < \infty,$$

and for $g: \mathbb{R} \to \mathbb{R}$,

$$E[g(X)] = \int_{-\infty}^{\infty} g(x)f(x)dx.$$

Cumulative Distribution Function (Continuous Case)

The cumulative distribution function, c.d.f., for X, denoted $F(x)$, is

$$F(x) = P(X \le x) = \int_{-\infty}^{x} f(t)dt$$

where $f(x)$ is the p.d.f. for X.

It follows that if $f(x)$ is continuous, then

$$\frac{d}{dx}F(x) = f(x).$$

Theorem 4.1

The cumulative distribution function F has the following properties:

(i) $F(x)$ is nondecreasing; that is, if $x < y$, then $F(x) \le F(y)$.
(ii) $\lim_{x\to-\infty} F(x) = 0$.
(iii) $\lim_{x\to\infty} F(x) = 1$.
(iv) $F(x)$ is continuous from the right; that is, $\lim_{x\downarrow x_0} F(x) = F(x_0)$.
(v) The left-hand limit of $F(x)$ exists.

Proof

NOTE: If the c.d.f. is as defined by Equation 4.1, then it is continuous and properties (iv) and (v) do not require a proof. In some cases, the random variable is a hybrid of the discrete and continuous cases, and the proof is given for those situations.

Parts of the proof rely on the continuity properties that were proven in Chapter 3. In particular,

$$\text{if } E_1 \subset E_2 \subset \cdots, \quad \text{then } P\left(\cup_{n=1}^{\infty} E_n\right) = \lim_{n\to\infty} P(E_n)$$

and

$$\text{if } E_1 \supset E_2 \supset \cdots, \quad \text{then } P\left(\cap_{n=1}^{\infty} E_n\right) = \lim_{n\to\infty} P(E_n).$$

(i) If $x < y$, then $F(y) - F(x) = \int_{-\infty}^{y} f(t)\,dt - \int_{-\infty}^{x} f(t)\,dt = \int_{x}^{y} f(t)\,dt \ge 0$.

(ii) Let $E_n = \{X \le -n\}$. Then $E_1 \supset E_2 \supset \cdots$ and $\cap_{n=1}^{\infty} E_n = \phi$, so

$$\lim_{x\to-\infty} F(x) = \lim_{n\to\infty} P(E_n) = P\left(\cap_{n=1}^{\infty} E_n\right) = \lim_{n\to\infty} P(\phi) = 0.$$

(iii) Let $E_n = \{X \le n\}$. Then $E_1 \subset E_2 \subset \cdots$, so

$$\lim_{x\to\infty} F(x) = \lim_{n\to\infty} P\left(\cup_{n=1}^{\infty} E_n\right) = P(\mathbb{R}) = 1.$$

(iv) Let $\{x_n\}$ be a decreasing sequence of real numbers with

$$\lim_{n\to\infty} x_n = x_0,$$

and let $E_n = \{X \le x_n\}$. Then $E_1 \supset E_2 \supset \cdots$ and $\cap_{n=1}^{\infty} E_n = \{X \le x_0\}$, so

$$F(x_0) = P(X \le x_0) = P\left(\cap_{n=1}^{\infty} E_n\right) = \lim_{n\to\infty} P(E_n) = \lim_{n\to\infty} F(x_n)$$

and thus $F(x)$ is continuous from the right.

(v) Let $\{x_n\}$ be an increasing sequence of real numbers with

$$\lim_{n\to\infty} x_n = x_0.$$

Let $E_n = \{X \le x_n\}$. Then $E_1 \subset E_2 \subset \cdots$, and $\cup_{n=1}^{\infty} E_n = \{X < x_0\}$, so $\lim_{n\to\infty} P\left(\cup_{n=1}^{\infty} E_n\right) = \lim_{n\to\infty} P(E_n)$. Now $\{P(E_n)\}$ is an increasing sequence of numbers bounded earlier by 1, so

$$\lim_{x\uparrow x_0} F(x) = P(X < x_0) = \lim_{n\to\infty} P(E_n)$$

exists.

Exponential Distribution

The exponential distribution is a continuous distribution that is fundamental to the Poisson process, which is the simplest example of a continuous Markov chain.

Definition

The function

$$f(x) = \begin{cases} \lambda e^{-\lambda x}, & x \ge 0 \\ 0, & x < 0 \end{cases}$$

is a p.d.f. for any $\lambda > 0$. This probability distribution is the exponential distribution with parameter λ and a continuous random variable with such a function as its p.d.f. is an exponential random variable with parameter λ.

For an exponential random variable with parameter λ,

$$E\left[X^n\right] = \int_0^\infty x^n \lambda e^{-\lambda x} dx = \frac{n!}{\lambda^n}$$

so

$$E[X]=\frac{1}{\lambda},\quad E\left[X^2\right]=\frac{2}{\lambda^2},\quad \text{and}\quad \mathrm{Var}[X]=E\left[X^2\right]-\left(E\left[X\right]\right)^2=\frac{1}{\lambda^2}.$$

We also have the cumulative density function for an exponential random variable with parameter λ is

$$F(x)=\int_0^x \lambda e^{-\lambda t}dt=-e^{-\lambda t}\Big|_0^x=1-e^{-\lambda x}.$$

We will often use the fact that if X is an exponential random variable with parameter λ, then for $x > 0$

$$P(X > x)=e^{-\lambda x}.$$

The exponential distribution is commonly used to determine how likely is it that some event will occur in a given period of time.

Example

In this type of problem, λ gives the rate of occurrence and so 1/λ gives the average wait between occurrences. So if we expect 4 customers an hour to enter a store, then the rate of occurrence is λ = 4 per hour, and the average wait between customers is $1/\lambda = 1/4$ h or 15 min.

Suppose that the amount of time spent on a task is exponentially distributed, with a mean of 20 min. What is the probability that a task will take between 25 and 35 min to accomplish?

Solution

The mean of an exponential distribution is 1/λ, so here we are given that 1/λ = 20 and so λ = .05. Thus,

$$P(25 < x < 35)=\int_{25}^{35} .05e^{-.05x}dx \approx .113.$$

In many applications of the exponential function, we are given the rate at which some event occurs. The event might be that a customer arrives or an electrical component malfunctions. A typical question would be, what is the probability that a customer arrives in a certain period of time? It has been shown that in a large number of such scenarios, the situation is accurately modeled by an exponential distribution (see, e.g., Feller 1968). In the following, we show why, under certain hypotheses, the exponential function is the appropriate distribution for such a model, but we first give some necessary background.

$o(h)$ Functions

In formulating the equations for the transition probabilities of customer arrivals, we will use the notion of $o(h)$.

Definition

A function $f(x)$ is $o(h)$ if

$$\lim_{h \to 0} \frac{f(h)}{h} = 0.$$

If $f(x)$ and $g(x)$ are $o(h)$, then $f(x) + g(x)$ is $o(h)$, and if $f(x)$ is $o(h)$ and $h(x)$ is bounded, then $f(x)h(x)$ is $o(h)$.

One way to determine if a function is $o(h)$ is by its Maclaurin series. If the lowest power of h in the Maclaurin series for the function is two or larger, then the function is $o(h)$. An example that we will often use is

$$1 - e^{-ah} = 1 - \left(1 - ah + \frac{(ah)^2}{2!} + \cdots\right) = ah + o(h).$$

Exponential Distribution as a Model for Arrivals

We suppose that the arrival of customers follows the following conditions:

1. The customers arrive at a constant rate, which we denote λ.
2. The arrival of customers over one time interval does not affect the arrival of customers over a disjoint time interval.
3. Customers arrive one at a time.

Let X_t be the random variable that gives the number of arrivals up to and including time t. For a small positive number h, these conditions can be modeled by the following equations:

(a) By condition 1, the probability that one customer arrives in the interval $[t, t + h]$ is $\lambda h + o(h)$. This can be stated as

$$P(X_{t+h} = X_t + 1) = \lambda h + o(h).$$

(b) Condition 3 can be stated as

$$P\{X_{t+h} \geq X_t + 2\} = o(h).$$

(c) Together, these equations imply

$$P(X_{t+h} = X_t) = 1 - \lambda h + o(h).$$

Theorem 4.2

Let X_t be the random variable that gives the number of arrivals up to and including time t, and suppose that the arrivals satisfy conditions 1–3.

Let $P_k(t)$ be the probability that there have been exactly k arrivals by time t; that is,

$$P_k(t) = \Pr\{X_t = k\}, \quad k \text{ a nonnegative integer.}$$

Then

$$P_k(t) = \frac{(\lambda t)^k}{k!} e^{-\lambda t}.$$

Proof

We seek a formula for

$$\frac{d}{dt} P_k(t).$$

Now

$$\frac{d}{dt} P_k(t) = \lim_{h \to 0} \frac{P_k(t+h) - P_k(t)}{h}$$

and

$$P_k(t+h) = \Pr(X_{t+h} = k).$$

We want to know how could $X_{t+h} = k$ knowing what X_t is.

1. One way is if $X_t = k$ and no change occurs between t and $t + h$. The probability that no change occurs between t and $t + h$ is $1 - \lambda h + o(h)$. The probability of both events occurring together is

$$P_k(t)(1-\lambda h+o(h)).$$

2. Another possibility is if $X_t = k - 1$ and exactly one change occurs between t and $t + h$. The probability that exactly one change occurs between t and $t + h$ is $\lambda h + o(h)$. The probability of both events occurring together is

$$P_{k-1}(t)(\lambda h+o(h)).$$

3. The last possibility is if $X_t = k - n, n \geq 2$ and n changes occur between t and $t + h$. The probability of two or more changes occurring between t and $t + h$ is $o(h)$.

Thus,

$$P_k(t+h) = P_k(t)(1-\lambda h+o(h)) + P_{k-1}(t)(\lambda h+o(h)) + o(h).$$

Then

$$P_k(t+h) - P_k(t) = P_k(t)(-\lambda h) + P_{k-1}(t)(\lambda h) + o(h)$$

and

$$\frac{P_k(t+h)-P_k(t)}{h} = \lambda P_{k-1}(t) - \lambda P_k(t) + \frac{o(h)}{h}.$$

Finally, we have

$$P_k'(t) = \lim_{h\to 0}\frac{P_k(t+h)-P_k(t)}{h} = \lim_{h\to 0}\left[\lambda P_{k-1}(t) - \lambda P_k(t) + \frac{o(h)}{h}\right]$$

$$= \lambda P_{k-1}(t) - \lambda P_k(t).$$

Now define $P_{-1}(t) = 0$ and we have

$$P_0'(t) = -\lambda P_0(t)$$

so

$$P_0(t) = Ce^{-\lambda t}.$$

Since $P_0(0) = 1$, we have $C = 1$ and so

$$P_0(t) = e^{-\lambda t}.$$

For $k = 1$, we have

$$P_1'(t) = \lambda P_0(t) - \lambda P_1(t)$$

so

$$P_1'(t) + \lambda P_1(t) = \lambda P_0(t) = \lambda e^{-\lambda t}. \tag{1}$$

With $y = P_1(t)$, Equation 1 is of the form

$$y' + \lambda y = \lambda e^{-\lambda t}$$

and we use the integrating factor $e^{\lambda t}$ to get

$$e^{\lambda t}(y' + \lambda y) = e^{\lambda t}\lambda e^{-\lambda t} = \lambda.$$

Now

$$\frac{d}{dt}\left(ye^{\lambda t}\right) = e^{\lambda t}(y' + \lambda y) = \lambda$$

so

$$ye^{\lambda t} = \lambda t + C \quad \text{and} \quad y = e^{-\lambda t}(\lambda t + C)$$

and

$$P_1(t) = y = e^{-\lambda t}(\lambda t + C).$$

Since $P_1(0) = 0$, we have $C = 0$ and

$$P_1(t) = e^{-\lambda t}\lambda t.$$

We seek a pattern for $P_n(t)$. For $n = 2$, we have

$$P_2'(t) = \lambda P_1(t) - \lambda P_2(t)$$

so

$$P_2'(t) + \lambda P_2(t) = \lambda P_1(t) = \lambda\left(e^{-\lambda t}\lambda t\right).$$

Letting $y = P_2(t)$, this is of the form

$$y' + \lambda y = \lambda^2 t e^{-\lambda t}.$$

Again using the integrating factor $e^{\lambda t}$, we obtain

$$e^{\lambda t}(y' + \lambda y) = e^{\lambda t}\lambda^2 t e^{-\lambda t} = \lambda^2 t$$

and

$$\frac{d}{dt}\left(ye^{\lambda t}\right) = e^{\lambda t}(y' + \lambda y) = \lambda^2 t$$

so

$$ye^{\lambda t} = \frac{\lambda^2 t^2}{2} + C$$

and

$$P_2(t) = y = e^{-\lambda t}\left(\frac{\lambda^2 t^2}{2} + C\right).$$

Since $P_2(0) = 0$, we have $C = 0$ and

$$P_2(t) = e^{-\lambda t}\frac{\lambda^2 t^2}{2}.$$

A pattern is beginning to suggest itself, namely,

$$P_k(t) = e^{-\lambda t}\frac{\lambda^k t^k}{k!}.$$

In Exercise (15), we complete the proof of the theorem by showing that this is indeed the case.

Corollary

For processes that obey conditions 1 and 3, the probability of an arrival has an exponential distribution.

Proof

Let X be the random variable that there has been no arrival. We have

$$F(t) = P(X < t) = 1 - e^{-\lambda t}$$

so

$$f(t) = \frac{dF}{dt} = \lambda e^{-\lambda t}.$$

The simplest example of a continuous Markov process is a Poisson process, and its construction depends on the exponential distribution. In the Poisson process, we want to count the number of occurrences of some event in a specified unit of time, similar to the renewal theory. We assume that the time between successive occurrences of the event follows an exponential distribution and that the times between successive occurrences are independent and identically distributed.

One connection of these ideas is that all deal with the waiting time before some event occurs. Another is the memoryless property, which is necessary for the Markov property. We will show that the exponential distribution is the only continuous probability distribution that has the memoryless property.

Memoryless Random Variables

In continuous Markov processes, the characteristic that is analogous to the Markov property in discrete chains is that the random variables be memoryless.

Definition

A random variable X that satisfies

$$P(X > t + s | X > s) = P(X > t)$$

is said to lack memory, or be memoryless.

Our next two theorems prove the crucial property that a continuous probability distribution is memoryless if and only if it is an exponential distribution. In the discrete case, the geometric distribution is memoryless.

Theorem 4.3

If X is an exponential random variable, then $P(X > s + t|X > s) = P(X > t)$ for $t > 0$ and $s \geq 0$. Thus, an exponential random variable is memoryless.

Proof

We have

$$P(X > s+t \mid X > s) = \frac{P(X > s+t, X > s)}{P(X > s)} = \frac{P(X > s+t)}{P(X > s)} = \frac{e^{-\lambda(s+t)}}{e^{-\lambda s}}$$
$$= e^{-\lambda t} = P(X > t).$$

Corollary

If X is an exponential random variable, then $P(X > s + t) = P(X > t)P(X > s)$.

Proof

We have

$$P(X > s+t) = e^{-\lambda(s+t)} = e^{-\lambda(t)}e^{-\lambda(s)} = P(X > t)P(X > s).$$

We now show that continuous memoryless random variables have an exponential distribution.

Theorem 4.4

If $f(x)$ is the p.d.f. of a continuous random variable that is memoryless, then $f(x)$ is an exponential function.

Proof

Suppose that X is a continuous random variable with p.d.f. $f(x)$ and c.d.f. $F(x)$ and X is memoryless. Let

$$G(t) = 1 - F(t) = P(X > t).$$

Note that $G(t)$ is monotone decreasing.

Since X is memoryless, we have

$$P(X > t+s \mid X > t) = P(X > s).$$

Now

$$P(X > t+s \mid X > t) = \frac{P(X > t+s, X > t)}{P(X > t)} = \frac{P(X > t+s)}{P(X > t)}$$

so

$$\frac{P(X > t+s)}{P(X > t)} = P(X > s)$$

and

$$P(X > t+s) = P(X > s)P(X > t).$$

Thus, we have

$$G(t+s) = G(t)G(s). \tag{2}$$

We next show that if G is a continuous function defined for $t > 0$ that satisfies Equation 2, then $G(t) = 0$ or there is a constant $\lambda > 0$ for which

$$G(t) = e^{-\lambda t}.$$

Since $G(t) = 1 - F(t) = P(X > t)$ cannot be 0 for all $t > 0$. Thus, once we have proven the previous assertion, then we will have

$$G(t) = e^{-\lambda t}.$$

From Equation 2,

$$G(2) = G(1+1) = G(1)G(1) = [G(1)]^2$$

and it follows that

$$G(n) = [G(1)]^n.$$

Also

$$G(1) = G\left(\frac{n}{n}\right) = G\left(\frac{1}{n} + \cdots + \frac{1}{n}\right) = \left[G\left(\frac{1}{n}\right)\right]^n$$

so

$$G\left(\frac{1}{n}\right)=\left[G(1)\right]^{\frac{1}{n}}.$$

Then

$$G\left(\frac{m}{n}\right)=G\left(\frac{1}{n}+\cdots+\frac{1}{n}\right)=\left[G\left(\frac{1}{n}\right)\right]^{m}=\left[\left[G(1)\right]^{\frac{1}{n}}\right]^{m}=\left[G(1)\right]^{\frac{m}{n}}.$$

So for each positive rational number, we have

$$G\left(\frac{m}{n}\right)=\left[G(1)\right]^{\frac{m}{n}}.$$

Since $G(t)$ is monotone, it follows that

$$G(x)=\left[G(1)\right]^{x}$$

for all positive real numbers.

Now $\left[G(1)\right]^{x}=e^{-\lambda x}$, if $\lambda=-\ln\left[G(1)\right]$.

The c.d.f. for X is

$$F(x)=P(X\le x)=1-P(X>x)=1-G(x)=1-e^{-\lambda x}.$$

So $F(x)$ is differentiable and the p.d.f. of X is

$$\frac{d}{dx}F(X)=f(x)=\lambda e^{-\lambda x}$$

so X is an exponential random variable.

Thus, we have *a continuous random variable that is memoryless if and only if it has an exponential distribution.*

Because of the memoryless property, the exponential distribution can sometimes be counterintuitive. For example, suppose you are at a bank and there are two tellers, both of whom are busy and you have to select one line in which to wait. If you believe the tellers are equally adept, and you have no way of knowing the complexity of the transactions each is dealing with, and you know one teller has been with a customer for 10 min and the other has been with a customer for 3 min, most people would probably choose to wait for the teller who had been with the customer for 10 min.

As another example, if you know the average lifetime of a light bulb is 1000 h and you are offered a light bulb that has burned for 1200 h versus a new light bulb, most would probably take the new light bulb.

Despite the intuition of the previous examples, the exponential distribution has proven to be an excellent model in a multitude of settings. However, one must be sure the hypotheses of the experiment fit the distribution.

Example

The amount of time it takes for a clerk to serve a customer is exponentially distributed with a mean of 20 min.

(a) What is the probability that it will require more than 30 min to serve a customer?

Solution

Since the expected value is

$$E[X] = \frac{1}{\lambda} = 20, \quad \text{then } \lambda = .05$$

and

$$P(X > 30) = \int_{30}^{\infty} .05e^{-.05t}dt \approx .223.$$

(b) A customer has been with a clerk for 15 min. What is the probability she will be there at least 10 more?

Solution

We have

$$P(X > s + t \mid X > s) = P(X > t)$$

so

$$P(X > 15 + 10 \mid X > 15) = P(X > 10) = \int_{10}^{\infty} .05e^{-.05t}dt \approx .607.$$

(c) Suppose there is a store with two clerks and the amount of time it takes for each clerk to serve a customer is exponentially distributed with a mean of 20 min. Suppose further there are three customers A, B, and C. Customers A and B are being served and C is waiting in line. Customer A has been with a clerk for 10 min and customer B for 20 min. What can you say about the probability of each customer leaving the store first, second, and third?

Solution

By the memoryless property, A and B each finish before the other with a probability of 1/2. Since one of A or B must leave before C can be served, C cannot leave first.

If A leaves first, then B leaves before C with a probability of ½ and this is the only way B can leave second. The probability of B leaving second is the intersection of the events A leaves first and B finishes service before C. The probability of the intersection of these two independent events is

$$\frac{1}{2}\cdot\frac{1}{2}=\frac{1}{4}.$$

Since B leaves first with probability 1/2, and second with probability 1/4, then he leaves third with probability 1/4.

By a symmetric argument, the probabilities for A are the same as those for B.

We are left with the probability of C leaving second is 1/2 and the probability of C leaving third is 1/2.

Theorem 4.5

Suppose that $X_1, \ldots, X_n$ are independent exponential random variables with rates $\lambda_1, \ldots, \lambda_n$, respectively. Let $X = \min\{X_1, \ldots, X_n\}$.

(a) Then

$$P(X>t)=e^{-\lambda_1 t}\cdots e^{-\lambda_n t}=e^{-(\lambda_1+\cdots+\lambda_n)t}.$$

(b) We have

$$P(X=X_k)=\frac{\lambda_k}{\lambda_1+\cdots+\lambda_n}.$$

Proof

(a) Since $X_1, \ldots, X_n$ are independent and $X>t$ if and only if $X_k>t$ for every $k=1, \ldots, n$, we have

$$P(X>t)=P(X_1>t,\ldots,X_n>t)=P(X_1>t)\cdots P(X_n>t)$$
$$=e^{-\lambda_1 t}\cdots e^{-\lambda_n t}=e^{-(\lambda_1+\cdots+\lambda_n)t}.$$

(b) To demonstrate the proof, we do the case for $k = 1$. Since $X = X_1$ if and only if there is a t for which $X_1 = t$ and $X_j > t$ for $j = 2, \ldots, n$. Using the law of total probability, we have

$$P(X = X_1) = \int_0^\infty P(X_2 > t, X_3 > t, \ldots, X_n > t)\, dP(X_1 = t)$$

$$= \int_0^\infty e^{-(\lambda_2 + \cdots + \lambda_n)t} \lambda_1 e^{-\lambda_1 t}\, dt = \frac{\lambda_1}{\lambda_1 + \lambda_2 + \cdots + \lambda_n}.$$

We explain why

$$P(X = X_1) = \int_0^\infty P(X_2 > t, X_3 > t, \ldots, X_n > t)\, dP(X_1 = t) = \int_0^\infty e^{-(\lambda_2 + \cdots + \lambda_n)t} \lambda_1 e^{-\lambda_1 t}\, dt.$$

Fix $\Delta t > 0$. Divide the interval $[0, \infty)$ into intervals of width Δt. Now $P(X = X_1)$ if and only if there is a nonnegative integer k with

$$X_1 \in \left[k\Delta t, (k+1)\Delta t\right] \quad \text{and} \quad X_j > (k+1)\Delta t, \quad j = 2, \ldots, n.$$

Now

$$P\left(X_j > (k+1)\Delta t\right) = e^{-\lambda_j (k+1)\Delta t}$$

and

$$P\left(X_1 \in \left[k\Delta t, (k+1)\Delta t\right]\right) \approx \lambda_1 e^{-\lambda_1 k \Delta t} \Delta t.$$

Thus,

$$P(X = X_1) \approx \sum_{k=1}^{\infty} \left[P\left(X_1 \in \left[k\Delta t, (k+1)\Delta t\right]\right) \prod_{j=2}^{n} P\left(X_j > (k+1)\Delta t\right) \right]$$

$$\approx \sum_{k=1}^{\infty} \left[\lambda_1 e^{-\lambda_1 k \Delta t} \Delta t \prod_{j=2}^{n} e^{-\lambda_j (k+1)\Delta t} \right]. \tag{3}$$

Taking the limit as $\Delta t \to 0$ in (3) gives the exact answer, so

$$P(X = X_1) = \int_0^\infty \lambda_1 e^{-\lambda_1 t} \prod_{j=2}^n e^{-\lambda_j t} dt = \int_0^\infty P(X_2 > t, X_3 > t, \ldots, X_n > t) dP(X_1 = t).$$

In Exercise 4.10, we show that if $X_1, \ldots, X_n$ are independent exponential random variables with rates $\lambda_1, \ldots, \lambda_n$, respectively, and $X = \max\{X_1, \ldots, X_n\}$, then

$$P(X > t) = \prod_{i=1}^n \left(1 - e^{-\lambda_i t}\right).$$

Example

We have an electrical device that consists of three components, and the device fails if any of the components fails. Suppose the failure rate of each component is exponentially distributed and the failure rate of the ith component is

$$\lambda_i = \frac{1}{2^i}, \quad i = 1, 2, 3.$$

(a) What is the failure rate of the system, and what is the probability that each component will cause the failure?
We have

$$\lambda_1 + \lambda_2 + \lambda_3 = \frac{1}{2} + \frac{1}{4} + \frac{1}{8} = \frac{7}{8}$$

so if X is the random variable that gives the time of failure of the system, then

$$P(X > t) = e^{-(7/8)t}.$$

The probability that the ith component caused the failure is

$$\frac{\lambda_i}{\lambda_1 + \lambda_2 + \lambda_3} = \frac{\lambda_i}{(7/8)} = \frac{8\lambda_i}{7}.$$

So the probability that component 1, 2, or 3 caused the failure is

$$\frac{8(1/2)}{7} = \frac{4}{7}, \quad \frac{8(1/4)}{7} = \frac{2}{7}, \quad \text{and} \quad \frac{8(1/8)}{7} = \frac{1}{7},$$

respectively.

(b) Suppose that the time of repair depends on the component that failed and the time of repair for each component is exponentially distributed with parameter μ_i with $\mu_1 = \frac{1}{3}$, $\mu_2 = \frac{1}{10}$, $\mu_3 = 1$. What is the expected amount of time the system will be down in case of a failure?

We use that probability the system fails due to a failure of component i is

$$\frac{\lambda_i}{\lambda_1 + \lambda_2 + \lambda_3}$$

and the expected downtime if component i fails is

$$\frac{1}{\mu_i}.$$

Component 1:

$$\text{Probability of failure} = \frac{\lambda_1}{\lambda_1 + \lambda_2 + \lambda_3} = \frac{1/2}{7/8} = \frac{4}{7}$$

$$\text{expected length of outage} = \frac{1}{\mu_1} = 3.$$

Component 2:

$$\text{Probability of failure} = \frac{\lambda_2}{\lambda_1 + \lambda_2 + \lambda_3} = \frac{1/4}{7/8} = \frac{2}{7}$$

$$\text{expected length of outage} = \frac{1}{\mu_2} = 10.$$

Component 3:

$$\text{Probability of failure} = \frac{\lambda_1}{\lambda_1 + \lambda_2 + \lambda_3} = \frac{1/8}{7/8} = \frac{1}{7}$$

$$\text{expected length of outage} = \frac{1}{\mu_3} = 1.$$

The expected length of outage is

$$\sum_i (\text{Prob.comp.}i\,\text{fails})(\text{Expected length of outage for comp. } i)$$

$$= \left(\frac{4}{7}\right)(3) + \left(\frac{2}{7}\right)(10) + \left(\frac{1}{7}\right)(1) = \frac{33}{7}.$$

Poisson Process

A Poisson process is a counting process that satisfies a group of axioms. We want to determine the probability that a particular number of *events* occur within a certain period of time under certain assumptions. In order to make the ideas more intuitive, suppose that we want to count the arrival of customers during a period of time. We assume the following:

1. The arrival of customers during one time period does not affect the arrival of customers at a nonoverlapping time period.
2. The arrival rate is constant.
3. Customers arrive one at a time.

Note that conditions 2 and 3 are the same as those for the arrival of customers in the "Exponential Distribution" section. Thus, we know that the arrivals occur at an exponential rate.

In some instances, these assumptions may not be valid. For example, at a restaurant, the arrival rate of customers between 2 and 3 p.m. is almost certainly different from that between 6 and 7 p.m.

We formalize the aforementioned conditions to define a Poisson process.

Definition

Let N_t, $t \geq 0$ be a stochastic process where N_t is the number of times that some event has occurred up to and including time t. Suppose that N_t satisfies the following axioms:

1. $N_0 = 0$.
2. For times

$$0 \leq s_1 < t_1 \leq s_2 < t_2 \leq \cdots \leq s_n < t_n,$$

the random variables Y_k defined by

$$Y_k = N_{t_k} - N_{s_k}, \quad k = 1, \ldots, n$$

are independent and stationary. (Stationary means the distribution of $N_{t_2} - N_{t_1}$ has the same distribution as $N_{t_2+s} - N_{t_1+s}$ for any $0 \leq t_1 \leq t_2$ and any $s \geq 0$.)

3. There is a number $\lambda > 0$ for which

$$P(N_{t+h} = N_t) = 1 - \lambda h + o(h)$$

$$P(N_{t+h} = N_t + 1) = \lambda h + o(h)$$

$$P(N_{t+h} \geq N_t + 2) = o(h)$$

independent of t.

Then N_t is a Poisson process with parameter λ.

In Theorem 4.2, we showed that if X_t is a random variable that satisfied

$$P(X_{t+h} = X_t) = 1 - \lambda h + o(h)$$

$$P(X_{t+h} = X_t + 1) = \lambda h + o(h)$$

$$P\{X_{t+h} \geq X_t + 2\} = o(h),$$

then

$$\Pr\{X_t = k\} = \frac{(\lambda t)^k}{k!} e^{-\lambda t}, \quad k \text{ being a nonnegative integer.}$$

This says, in particular, that

$$P(N_t = 0) = e^{-\lambda t}.$$

The most basic question involving a Poisson process is as follows: If we have a group of occurrences that satisfy the axioms of a Poisson process, how many occurrences have we had up to time t?

The initial part of the development for the Poisson process is very much like that of renewal processes. The setting is that some event (such as a customer arriving) occurs at random times, and we keep track of the arrival times.

There are other collections of random variables associated with a Poisson process. Let

$$X_1 = \text{the time of first arrival and}$$

$$X_n = \text{the time between the } (n-1)\text{st and } n\text{th arrivals.}$$

So $\{X_n\}$ is a sequence of random variables that give the time between consecutive arrivals.

The next theorem says that if we begin observing a Poisson process at some time $t_0 > 0$, then probabilistically it is the same as if we began observing the process at $t = 0$.

Theorem 4.6

For a Poisson process of rate λ, the interarrival times $\{X_n\}$ are independent, identically distributed exponential random variables with parameter λ.

Proof

Independence is a consequence of the second axiom in the definition of a Poisson process.

In Theorem 4.2, we showed that if X_t is a random variable that satisfied

$$P(X_{t+h} = X_t) = 1 - \lambda h + o(h),$$

$$P(X_{t+h} = X_t + 1) = \lambda h + o(h), \quad \text{and}$$

$$P\{X_{t+h} \geq X_t + 2\} = o(h),$$

then

$$\mathrm{P}\{X_t = k\} = \frac{(\lambda t)^k}{k!} e^{-\lambda t}, \quad k \text{ being a nonnegative integer.}$$

This says, in particular, that

$$P(N_t = 0) = e^{-\lambda t}$$

so X_1 is exponentially distributed with parameter λ.

To show X_2 is exponentially distributed with parameter λ, consider

$$\begin{aligned} P(X_2 > t | X_1 = s) &= P(0 \text{ arrivals in } (s, s+t] | X_1 = s) \\ &= P(0 \text{ arrivals in } (s, s+t]) \quad (\text{by independence}) \\ &= P(0 \text{ arrivals in } (0, t]) \quad (\text{by stationarity}) \\ &= e^{-\lambda t}. \end{aligned}$$

Showing that X_n, $n > 2$, is exponentially distributed with parameter λ is done in a similar manner.

Define the sequence of random variables $\{T_n\}$ by

$$T_n = \sum_{k=1}^{n} X_k.$$

So T_n is the time of the nth arrival. Note that $\{X_n\}$ can be recovered from $\{T_n\}$ and $\{T_n\}$ can be recovered from $\{X_n\}$.

Since, as with renewal processes, for

$$N_t = \text{the number of arrivals in}(0,t]$$

the graph of N_t is a step function that changes value at times $\{T_n\}$ and we again have the relations as in renewal processes,

$$T_n = \min_{t\geq 0}\{N_t = n\},\ \ N_t = \max_{n\in\mathbb{N}}\{T_n \leq t\},$$

and

$$N_t \geq n \quad \text{if and only if } T_n \leq t;\ n\in\mathbb{N},\ \ t\geq 0.$$

Theorem 4.7

Suppose that $X_1, X_2, \ldots, X_n$ are independent, identically distributed exponential random variables with parameter λ. Then the p.d.f. of $X_1 + X_2 + \cdots + X_n$ is

$$f_{X_1+X_2+\cdots+X_n}(t) = \lambda e^{-\lambda t}\frac{(\lambda t)^{n-1}}{(n-1)!}.$$

This is the gamma distribution with rate parameter λ and shape parameter n.

Proof

We proceed by induction. The result is true for $n = 1$. Assume the result holds for $n = k$. For $n = k + 1$, we use

$$\begin{aligned}
f_{X_1+\cdots+X_k+X_{k+1}}(t) &= \left(f_{X_1+\cdots+X_k} * f_{X_{k+1}}\right)(t)\\
&= \int_0^\infty f_{X_{k+1}}(t-s)\, f_{X_1+\cdots+X_k}(s)\,ds\\
&= \int_0^t \lambda e^{-\lambda(t-s)}\lambda e^{-\lambda s}\frac{(\lambda s)^{k-1}}{(k-1)!}ds = \lambda^2 e^{-\lambda t}\lambda^{k-1}\frac{1}{(k-1)!}\int_0^t s^{k-1}ds\\
&= \lambda^{k+1}e^{-\lambda t}\frac{1}{(k-1)!}\frac{t^{k-1}}{(k)} = \lambda e^{-\lambda t}\frac{(\lambda t)^k}{(k)!}.
\end{aligned}$$

So the result is true for $n = k + 1$ and thus holds for all positive integers.

Corollary

If X_i is an exponential random variable with parameter λ and N_t and T_n are as previously stated, then

$$P(N_t = n) = e^{-\lambda t}\frac{(\lambda t)^n}{n!}.$$

Proof

Since

$$N_t \geq n \quad \text{if and only if } T_n \leq t,$$

we have

$$\begin{aligned} P(N_t \geq n) &= P(T_n \leq t) = P(X_1 + X_2 + \cdots + X_n \leq t) \\ &= \int_0^t f_{X_1+X_2+\cdots+X_n}(s)\,ds = \int_0^t \lambda e^{-\lambda s}\frac{(\lambda s)^{n-1}}{(n-1)!}\,ds. \end{aligned}$$

Using induction, it can be shown that

$$\int_0^t \lambda e^{-\lambda s}\frac{(\lambda s)^{n-1}}{(n-1)!}\,ds = 1 - \sum_{k=0}^{n-1} e^{-\lambda t}\frac{(\lambda t)^k}{k!}.$$

But

$$\begin{aligned} P(N_t = n) &= P(N_t \geq n) - P(N_t \geq n+1) \\ &= \left[1 - \sum_{k=0}^{n-1} e^{-\lambda t}\frac{(\lambda t)^k}{k!}\right] - \left[1 - \sum_{k=0}^{n} e^{-\lambda t}\frac{(\lambda t)^k}{k!}\right] = e^{-\lambda t}\frac{(\lambda t)^n}{n!}. \end{aligned}$$

Definition

A Poisson distribution is a probability distribution on the nonnegative integers given by

$$P(X = n) = e^{-\lambda}\frac{\lambda^n}{n!}, \quad \lambda > 0, n = 0,1,2,\ldots$$

where X gives the number of occurrences of some event in one unit of time.

This of course is what we have from the corollary by setting $t = 1$.

Example

Suppose there is a store where customers arrive on average 4 per hour. Assuming the hypotheses for a Poisson process are satisfied, what is the probability that

(a) Two or fewer customers arrive in an hour?
(b) Between three and six customers arrive in an hour?
(c) Five or more customers arrive in an hour?

Solution

(a) We have $\lambda = 4$, so we compute

$$e^{-4}\sum_{n=0}^{2}\frac{4^n}{n!} \approx .238.$$

(b) The probability that between three and six customers will arrive in an hour is

$$\sum_{n=3}^{6} e^{-4}\frac{4^n}{n!} \approx .651.$$

(c) The probability that five or more customers will arrive in an hour is

$$1 - e^{-4}\sum_{n=0}^{4}\frac{4^n}{n!} \approx 1 - .629 = .371.$$

Poisson Processes with Occurrences of Two Types

Consider the case where we have a Poisson process having rate λ and the occurrence of an event can be one of two types. An example would be customers entering a store. Some will be male and the others will be female. Suppose the probability the customer is a male is p and the probability the customer is a female is $1 - p$.

Let

$N(t)$ = the number of customers that enter the store in the interval of time $[0, t]$

$M(t)$ = the number of male customers that enter the store in the interval of time $[0, t]$

$F(t)$ = the number of female customers that enter the store in the interval of time $[0, t]$.

For a given value of t, the probability that exactly k customers enter the store in the interval of time $[0, t]$ is

$$P\left(N(t)=k\right)=e^{-\lambda t}\frac{(\lambda t)^k}{k!}.$$

We want to compute the probability that in the period $[0, t]$, exactly m male customers and n female customers have entered the store. To do this, we compute

$$P\left(M(t)=m,F(t)=n\right).$$

Using

$$P(A\cap B)=P(A\,|\,B)P(B),$$

we have

$$\begin{aligned}P\left(M(t)=m,F(t)=n\right)&=P\left(M(t)=m,N(t)=m+n\right)\\&=P\left(M(t)=m\,|\,N(t)=m+n\right)P\left(N(t)=m+n\right).\end{aligned}$$

Now $P\left(M(t)=m\,|\,N(t)=m+n\right)$ is a binomial distribution with $m + n$ trials and the probability of success on a given trial is p. Thus,

$$P\left(M(t)=m\,|\,N(t)=m+n\right)=\binom{m+n}{m}p^m(1-p)^n.$$

Also,

$$P\left(N(t)=m+n\right)=e^{-\lambda t}\frac{(\lambda t)^{(m+n)}}{(m+n)!}$$

so

$$P(M(t)=m,F(t)=n)=\binom{m+n}{m}p^m(1-p)^n e^{-\lambda t}\frac{(\lambda t)^{(m+n)}}{(m+n)!}$$

$$=\frac{(m+n)!}{m!n!}e^{-\lambda t}p^m(1-p)^n(\lambda t)^n\frac{1}{(m+n)!}$$

$$=e^{-p\lambda t}\frac{(p\lambda t)^m}{m!}e^{-(1-p)\lambda t}\frac{[(1-p)\lambda t]^n}{n!}$$

$$=\left\{e^{-p\lambda t}\frac{(p\lambda t)^m}{m!}\right\}\left\{e^{-(1-p)\lambda t}\frac{[(1-p)\lambda t]^n}{n!}\right\}$$

$$=P(M(t)=m)P(F(t)=n).$$

Note that this also shows that $M(t)$ and $F(t)$ are independent events.

Example

This example shows that the conditional distribution of a Poisson process is a binomial distribution. Suppose that the rate of the Poisson process is 5. We show that the conditional distribution for N_4 given that $N_9 = 7$ is binomial.

First, note that $P(N_4 = k|N_9 = 7) = 0$ unless $0 \le k \le 7$. In this case,

$$P(N_4=k\,|\,N_9=7)=\frac{P(N_4=k,N_9=7)}{P(N_9=7)}.$$

Now $N_4 = k$ and $N_9 = 7$ if and only if $N_9 - N_4 = 7 - k$, and since $N_0 = 0$, then $N_4 = k$ if and only if $N_4 - N_0 = k$ and $N_9 = 7$ if and only if $N_9 - N_0 = 7$. Thus,

$$\frac{P(N_4=k,N_9=7)}{P(N_9=7)}=\frac{P(N_4-N_0=k,N_9-N_4=7-k)}{P(N_9-N_0=7)}.$$

Now

$$P(N_4-N_0=k)=e^{-(5\cdot4)}\frac{(5\cdot4)^k}{k!}$$

$$P(N_9-N_4=7-k)=e^{-(5\cdot5)}\frac{(5\cdot5)^{(7-k)}}{(7-k)!}$$

$$P(N_9-N_0=7)=e^{-(5\cdot9)}\frac{(5\cdot9)^7}{7!}$$

so

$$P(N_4 = k \mid N_9 = 7) = \frac{e^{-(5\cdot 4)}\dfrac{(5\cdot 4)^k}{k!}e^{-(5\cdot 5)}\dfrac{(5\cdot 5)^{(7-k)}}{(7-k)!}}{e^{-(5\cdot 9)}\dfrac{(5\cdot 9)^7}{7!}}$$

$$= \frac{e^{-20}e^{-25}}{e^{-45}} \cdot \frac{\dfrac{20^k 25^{(7-k)}}{k!(7-k)!}}{\dfrac{(5\cdot 9)^7}{7!}} = \frac{7!}{k!(7-k)!} \cdot \frac{5^k 4^k 5^{(7-k)} 5^{(7-k)}}{5^7 9^7}$$

$$= \binom{7}{k} \cdot \frac{4^k 5^{(7-k)}}{9^7} = \binom{7}{k}\left(\frac{4}{9}\right)^k \left(\frac{5}{9}\right)^{7-k},$$

which is the p.d.f. for a binomial distribution.

In Exercise 4.13, we extend this example.

Exercises

4.1 Suppose that we have two Poisson processes, the first with rate λ_1 and the second with rate λ_2. We want to determine the probability that there are m occurrences of the first process before there are n occurrences of the second process.

(a) What is the probability that there is 1 occurrence of the first process before there are any occurrences of the second process?

(b) What is the probability that there are 2 occurrences of the first process before there are any occurrences of the second process?

(c) What is the probability that in the first 8 occurrences 5 will be from the first process and 3 from the second process?

(d) If we flip a coin that lands heads with probability p and tails with probability $q = 1 - p$, what is the probability we will get 4 heads before we get 2 tails?

(e) What is the probability that there are m occurrences of the first Poisson process before there are n occurrences of the second Poisson process?

4.2 A computer printer fails an average of 6 times in a 30-day month with a Poisson distribution.

(a) What is the probability that there are 2 or fewer failures in the first 15 days?

(b) What is the probability there are more than 4 failures in a given 10 day period?

(c) What is the probability that there will be between 2 and 5 failures in a given 8-day period?

(d) What is the probability that exactly 2 failed in the first week given that 5 failed in the first 3 weeks?

4.3 An archer shoots at a target in accordance with a Poisson distribution at a rate of 2 arrows per minute. He hits the bull's-eye with probability .1. What is the probability he hits the bull's-eye exactly 3 times in 5 min?

4.4 In a mathematics book, there are an average of 3 letter typos per page and 2 number typos per page, both types having a Poisson distribution.

(a) What is the distribution for the total number of errors per page?

(b) On a given page, what is the probability that there are exactly 2 errors of each type?

4.5 At a pizza shop, there are three kinds of pizza, ham, and mushroom. Orders arrive according to a Poisson distribution at the rate of 8 per hour, and the types of pizza are equally popular. What is the probability that there are exactly 2 ham pizzas ordered in the next hour?

4.6 The help desk at a computer services business gets an average of 12 calls per hour that are exponentially distributed.

(a) What is the probability that there are fewer than 8 calls between 8:15 and 8:45 a.m.?

(b) What is the probability that you will receive exactly 2 calls the first 15 min of your shift?

(c) If you have to take a 10 min bathroom break, what is the probability you will miss a call?

(d) How long do you have to wait on your shift to have a probability of .95 that you will receive at least 1 call?

(e) What is the probability that there will be exactly 3 calls in a 15 min period?

4.7 Suppose that in a certain precinct, 60% of the voters are Republican and 40% are Democrat. Assume that on the average there are 200 voters per hour at a certain polling station. What is the probability that on a given hour there are 100 Republican voters at that voting station?

4.8 Suppose that at a Mexican restaurant the customers are Hispanic with probability 2/3 and Anglo with probability 1/3. Suppose that customers arrive at a Poisson rate of 20 customers per hour. What is the probability that exactly 5 Anglo customers arrive in a given hour?

4.9 We have two independent Poisson processes, the first with rate λ_1 and the second with rate λ_2.

(a) What is the probability that an event from process 1 occurs before an event from process 2?

(b) What is the probability that two events from process 1 occur before an event from process 2?

(c) Let A represent an occurrence of an event from process 1 and B an occurrence of an event from process 2. What is the probability of the sequence of occurrences AABABB?

(d) If there are 5 occurrences from the two processes combined, what is the probability that exactly 3 were from process 1?

4.10 We have two machines that operate independently. The first fails at a Poisson rate of 4 times per year and the second fails at a Poisson rate of 7 times per year. What is the probability that the first machine fails twice before the second machine fails three times?

4.11 Show that if that $X_1, \ldots, X_n$ are independent exponential random variables with rates $\lambda_1, \ldots, \lambda_n$, respectively, and $X = \max\{X_1, \ldots, X_n\}$, then

$$P(X > t) = \prod_{i=1}^{n} \left(1 - e^{-\lambda_i t}\right).$$

4.12 In a Poisson process with rate λ, for $0 \le s < t$, find the distribution of N_{t-s} and the distribution of $N_t - N_s$.

4.13 Show that for a Poisson process, if $s < t$, then

$$P\left(N_s = k \mid N_t = n\right) = \binom{n}{k}\left(\frac{s}{t}\right)^k \left(1 - \frac{s}{t}\right)^{n-k}.$$

4.14 Show that if $N_1(t)$ is a Poisson process with parameter λ_1 and $N_2(t)$ is a Poisson process with parameter λ_2, then $N_1(t) + N_2(t)$ is a Poisson process with parameter $\lambda_1 + \lambda_2$.

4.15 Use induction to complete the proof of Theorem 6.2. That is, under the hypotheses of the theorem show that

$$P_k(t) = \frac{(\lambda t)^k}{k!} e^{-\lambda t}$$

for every positive integer k.

5

Continuous-Time Markov Chains

Introduction

In discrete-time Markov chains, a change of state could occur only at specific points in time. For continuous-time Markov chains, a change of state can occur at any time. We again only consider processes that are homogeneous in time. The state space is countable (here, this includes finite), and there will be some differences between the finite and infinite state cases.

One way to view a continuous-time Markov chain is to embed a certain type of discrete-time Markov chain into a process that changes at random times rather than fixed times. We consider what restriction is necessary on the discrete-time Markov chain in this approach.

Let P be the transition matrix that governs how the states change (but not when).

First, consider a discrete-time Markov chain as an observer would see it.

The observer knows that the process makes a decision about a transition at (and only at) the times $t = 1, 2, \ldots$. If the observer watches the process from $t = 1.5$ to $t = 2.5$ and the system is in state i throughout those times, she knows that at $t = 2$ the process made a decision and decided to remain in state i. Thus, it must be that $P(i, i) > 0$.

Now consider a continuous-time Markov chain as an observer would see it.

The observer can only tell when a change in state occurred and what that change was. If there was a state i for which $P(i, i) > 0$, then it would be possible that when the system is in state i, an *alarm* could go off, telling the system to change but, in fact, the system *decided* to stay in state i. In this case, the observer would not see any change, nor would she be aware that the system had made a decision. We want to avoid this ambiguity, so we require that the embedded chain will have $P(i, i) = 0$ for every state i.

Our construction of a continuous-time Markov process will involve combining a discrete Markov chain with $P(i, i) = 0$ and a random process that chooses when a change of state occurs.

To build a mathematical model of this process, we need to postulate a mechanism to determine when a change occurs, and the embedded Markov chain will provide a probability distribution that describes how the new state is selected knowing the present state.

In a continuous-time Markov chain, for each time $t \in [0, \infty)$, X_t is a random variable that gives the state of the system at time t. For continuous time, the Markov property is

$$P\left(X_t = i \mid X_r, \quad 0 \le r \le s\right) = P\left(X_t = x \mid X_s\right)$$

and the property of time homogeneity is

$$P = \left(X_{t+s} = i \mid X_s = j\right) P = \left(X_t = i \mid X_0 = j\right).$$

We first prove the mechanism that controls when a change of state occurs is governed by exponential random variables by showing that a continuous-time Markov process is memoryless.

NOTE: We will often refer to a continuous–time Markov process as simply a Markov process.

Theorem 5.1

Let T_i be the time that a time homogenous continuous Markov process spends in state i. Then for $s, t \geq 0$, we have

$$P\left(T_i > s + t \mid T_i > s\right) = P\left(T_i > t\right)$$

and so T_i is exponentially distributed and memoryless.

Proof

Suppose the process begins in state i. By time homogeneity, we may begin at $t = 0$.

The event $T_i > s$ is equivalent to the event $\{X_u = i, 0 \leq u \leq s\}$ and the event $T_i > s + t$ is equivalent to the event $\{X_u = i, 0 \leq u \leq s + t\}$. Thus,

$$\begin{aligned}
P\left(T_i > s + t \mid T_i > s\right) &= P\left(X_u = i, 0 \le u \le s + t \mid X_u = i, 0 \le u \le s\right) \\
&= P\left(X_u = i, s < u \le s + t \mid X_u = i, 0 \le u \le s\right) \\
&= P\left(X_u = i, s < u \le s + t \mid X_s = i\right) \quad (\text{by the Markov property}) \\
&= P\left(X_u = i, 0 < u \le t \mid X_0 = i\right) \quad (\text{by time homogeneity}) \\
&= P\left(T_i > t\right).
\end{aligned}$$

A major difference between the discrete- and continuous-time cases of Markov chains is the transition mechanism. In the continuous case, in addition to giving the probability that if the system is in state i, when a transition occurs the transition is to state j, we must also specify a way to determine when a transition occurs. Theorem 5.1 says that the transition mechanism must be linked to an exponential distribution because of the memoryless property.

One way to determine when a transition occurs is to let each state have its own *alarm clock* that goes off with a specified rate. Let λ_n denote the rate at which the alarm goes off when the process is in state n. Since the rate at which the alarms go off has memoryless property, if T_n is the time at which the alarm of the clock at state n goes off, then

$$P(T_n > t) = e^{-\lambda_n t}.$$

Recapping, in the case of continuous Markov chains, when formulating the equations for the transition from a state, we need to consider two factors. First, the parameter that determines the rate at which a transition occurs, and second, the probability distribution that gives the likelihood of the state to which the process changes. In fact, these two factors are independent.

There are different (equivalent) ways to describe the model. Here is the one that we will use.

Associated with each state x is a parameter $\lambda(x)$, which is the exponential rate at which the system changes out of state x. Let $p(x, y)$ denote the probability that when the system changes from state x, the change is to state y. So $x \neq y$ and

$$\sum_{y:y\neq x} p(x,y) = 1.$$

This enables us to create a stochastic matrix associated with the process, called the embedded matrix, namely,

$$P(x,y) = p(x,y).$$

Note that $P(x, x) = 0$ for all x.

Let $X(t)$ denote the state of the process at time t. We postulate

(a) For $t > s, P\big(X(t)\big|X(r), 0 \leq r \leq s\big) = P\big(X(t)\big|X(s)\big)$ Markov property

(b) For $s > 0, P\big(X(t+s)\big|X(s)\big) = P\big(X(t)\big|X(0)\big)$ time homogeneity

(c) For $h > 0$,

  (i) $P\big(X(t+h) = x\big|X(t) = x\big) = 1 - \lambda(x)h + o(h)$

  (ii) $P\big(X(t+h) = y\big|X(t) = x\big) = \lambda(x)p(x,y)h + o(h)$

  (iii) The probability that there is more than one transition in $[t, t+h]$ is $o(h)$.

We note that if $N(t)$ is the number of transitions up to and including time t, then

$$P\left(N(t+h)=n \middle| N(t)=n\right)=1-\lambda h+o(h)$$

$$P\left(N(t+h)=n+1 \middle| N(t)=n\right)=\lambda h+o(h)$$

$$P\left(N(t+h)>n+1 \middle| N(t)=n\right)=o(h).$$

Example

The simplest example of a continuous Markov chain is a Poisson process. Recall the postulates for a Poisson process where the event being counted is the arrival of a customer:

1. The arrival of customers during one time period does not affect the arrival of customers at a nonoverlapping time period.
2. The arrival rate is constant.
3. Customers arrive one at a time.
   Condition 1 means that the process is memoryless and hence has an exponential distribution. This, together with conditions 2 and 3, implies that

$$P\left(N(t+h)=n \middle| N(t)=n\right)=1-\lambda h+o(h)$$

$$P\left(N(t+h)=n+1 \middle| N(t)=n\right)=\lambda h+o(h)$$

$$P\left(N(t+h)>n+1 \middle| N(t)=n\right)=o(h).$$

Generators of Continuous Markov Chains: The Kolmogorov Forward and Backward Equations

A major goal for continuous Markov chains is to compute $P(X_t = j|X_0 = i)$ for all $t \geq 0$ and all states i and j. To do this, we let $P_{ij}(t) = P(X_t = j|X_0 = i)$, determine a differential equation that $P_{ij}(t)$ must satisfy, and then solve the equation. We let $\lambda(i)$ be the rate of transition out of state i and $p(i, j)$ be the probability of transition from state i to state j.
Consider

$$P'_{ij}(t)=\lim_{h\downarrow 0}\frac{P_{ij}(t+h)-P_{ij}(t)}{h}=\lim_{h\downarrow 0}\frac{P\left(X_{t+h}=j \middle| X_0=i\right)-P\left(X_t=j \middle| X_0=i\right)}{h}.$$

By the law of total probability,

$$P(X_{t+h} = j | X_0 = i) = P(X_{t+h} = j | X_t = j) P(X_t = j | X_0 = i) + \sum_{k:k \neq j} P(X_{t+h} = j \mid X_t = k) P(X_t = k | X_0 = i).$$

We have

$$P(X_{t+h} = j | X_t = j) P(X_t = j | X_0 = i) = [1 - \lambda(j)h + o(h)] P(X_t = j | X_0 = i) = [1 - \lambda(j)h + o(h)] P_{ij}(t),$$

and if $k \neq j$,

$$P(X_{t+h} = j | X_t = k) P(X_t = k | X_0 = i) = [\lambda(k) p(k,j) h + o(h)] P(X_t = k \mid X_0 = i) = [\lambda(k) p(k,j) h + o(h)] P_{ik}(t).$$

Thus,

$$P_{ij}(t+h) - P_{ij} = P(X_{t+h} = j | X_0 = i) - P(X_t = j | X_0 = i)$$

$$= \left\{ [1 - \lambda(j)h + o(h)] P_{ij}(t) + \sum_{k:k \neq j} [\lambda(k) p(k,j) h + o(h)] P_{ik}(t) \right\} - P_{ij}(t)$$

$$= [-\lambda(j)h + o(h)] P_{ij}(t) + \sum_{k:k \neq j} [\lambda(k) p(k,j) h + o(h)] P_{ik}(t)$$

and so

$$P'_{ij}(t) = \lim_{h \downarrow 0} \frac{P(X_{t+h} = j | X_0 = i) - P(X_t = j | X_0 = i)}{h}$$

$$= [-\lambda(j)] P_{ij}(t) + \sum_{k:k \neq j} [\lambda(k) p(k,j)] P_{ik}(t). \qquad (1)$$

NOTE: We have tacitly assumed that $\sum_{k:k \neq j} [\lambda(k) p(k,j) h + o(h)] P_{ik}(t)$ gives an expression of the form $\left[\sum_{k:k \neq j} [\lambda(k) p(k,j) h] P_{ik}(t) \right] + o(h)$. If the number of states is finite, there is no problem, but if there are an infinite number of states, this requires justification. We assume that this is valid for our models.

Equation (1) is a system of ordinary differential equations called the Kolmogorov forward differential equations. We now express that system of equations as a matrix equation.

Definition

Let G be the matrix

$$G_{ij} = \begin{cases} -\lambda(i) & \text{if } i = j \\ \lambda(i)p(i,j) & \text{if } i \neq j \end{cases}. \tag{2}$$

The matrix G is the infinitesimal generator for the Markov process whose change rates are $\lambda(i)$ and transition probabilities are $p(i, j)$.

This matrix G has the properties

$$G_{ij} \geq 0 \quad \text{if } i \neq j, \quad \text{and} \quad \sum_j G_{ii} = 0. \tag{3}$$

Now

$$P_tG = \begin{pmatrix} P_{11}(t) & P_{12}(t) & P_{13}(t) & \dots & P_{1n}(t) \\ P_{21}(t) & P_{22}(t) & P_{23}(t) & \dots & P_{2n}(t) \\ P_{31}(t) & P_{32}(t) & P_{33}(t) & \dots & P_{3n}(t) \\ \vdots & \vdots & \vdots & \dots & \vdots \\ P_{n1}(t) & P_{n2}(t) & P_{n3}(t) & \dots & P_{nn}(t) \end{pmatrix}$$

$$\times \begin{pmatrix} -\lambda(1) & \lambda(1)p(1,2) & \lambda(1)p(1,3) & \cdots & \lambda(i)p(i,j) \\ \lambda(2)p(2,1) & -\lambda(2) & \lambda(2)p(2,3) & \cdots & \lambda(i)p(i,j)(t) \\ \lambda(3)p(3,1) & \lambda(3)p(3,2) & -\lambda(3) & \cdots & P_{3n}(t) \\ \vdots & \vdots & \vdots & \cdots & \vdots \\ \lambda(n)p(n,1) & \lambda(n)p(n,2) & \lambda(n)p(n,3) & \cdots & -\lambda(n) \end{pmatrix}$$

and

$$\begin{aligned} (P_tG)_{ij} &= P_{i1}(t)\lambda(1)p(1,j) + P_{i2}(t)\lambda(2)p(2,j) + \cdots + P_{ij-1}(t)\lambda(j-1)p(j-1,j) \\ &\quad - P_{ij}(t)\lambda(j) + P_{ij+1}(t)\lambda(j+1)p(j+1,j) + \cdots + P_{in}(t)\lambda(n)p(n,j) \\ &= -P_{ij}(t)\lambda(j) + \sum_{k \neq j} P_{ik}(t)\lambda(k)p(k,j) = (P_t')_{ij}. \end{aligned}$$

Thus, we have the matrix equation

$$\frac{d}{dt}(P_t) = P_tG \tag{4}$$

for which the solution is

$$P_t = P_0 e^{tG} \tag{5}$$

where P_0 is the distribution of X_0.

Note the order of P_t and G in equation (4).

Next, we will study the Kolmogorov backward equations, and in matrix form, these will be expressed as

$$\frac{d}{dt}(P_t) = GP_t.$$

In the backward equations, we begin the process in state i at time $(-h)$ and determine differential equations that allow the process to be in state j at time t. We split the interval $[-h, t]$ into $[-h, 0]$ and $[0, t]$ and analyze how the transition could take place.

One way is that the process could remain in state i from $-h$ to 0, which occurs with probability $\left[1-\lambda(j)h+o(h)\right]$, and then move from state i at time 0 to state j at time t. This latter step occurs with probability $P_{ij}(t)$ Thus, the probability of this scenario is $\left[1-\lambda(i)h+o(h)\right]P_{ij}(t)$.

The other way is that the process could move from state i at time $-h$ to a different state k at time 0, which occurs with probability $\left[p(i,k)\lambda(j)h+o(h)\right]$, and then moves to state k at time t. This latter step occurs with probability $P_{kj}(t)$. Thus, the probability of this scenario is $\left[p(i,k)\lambda(j)h+o(h)\right]P_{kj}(t)$. This could happen for every $k \neq i$.

The idea of using $-h$ as a time was an artifact to make the reasoning easier to follow. The length of the interval is $t + h$ so we have

$$P_{ij}(t+h) = \left[1-\lambda(i)h+o(h)\right]P_{ij}(t) + \sum_{k:k\neq i}\left[p(i,k)\lambda(j)h+o(h)\right]P_{kj}(t).$$

Thus,

$$P_{ij}(t+h) - P_{ij}(t) = \left[-\lambda(i)h+o(h)\right]P_{ij}(t) + \sum_{k:k\neq i}\left[p(i,k)\lambda(j)h+o(h)\right]P_{kj}(t)$$

so

$$\lim_{h\to 0}\frac{P_{ij}(t+h)-P_{ij}(t)}{h} = \lim_{h\to 0}\left[\frac{-\lambda(i)h}{h}P_{ij}(t) + \frac{\sum_{k:k\neq i}\left[p(i,k)\lambda(j)h\right]P_{kj}(t)}{h} + \frac{o(h)}{h}\right]$$

and we have

$$P'_{ij}(t) = -\lambda(i)P_{ij}(t) + \sum_{k:k\neq i}\left[p(i,k)\lambda(j)\right]P_{kj}(t). \tag{6}$$

Equation (6) are the backward equations. These describe the probability that if the system is in state at j time $t + h$, then it was in state i at time 0. Compare these with the forward equations

$$P'_{ij}(t) = \left[-\lambda(j)\right]P_{ij}(t) + \sum_{k:k\neq j}\left[\lambda(k)p(k,j)\right]P_{ik}(t) \tag{1}$$

that describe the probability that if the system is in state i at time 0, then it is in state j at time $t + h$.

Earlier, we observed that the equations given in (1) could be written as the matrix equation

$$\frac{d}{dt}(P_t) = P_t G.$$

We now show that the equations given in (6) can be written as the matrix equation

$$\frac{d}{dt}(P_t) = GP_t.$$

We have

$$GP_t = \begin{pmatrix} -\lambda(1) & \lambda(1)p(1,2) & \lambda(1)p(1,3) & \cdots & \lambda(i)p(i,j) \\ \lambda(2)p(2,1) & -\lambda(2) & \lambda(2)p(2,3) & \cdots & \lambda(i)p(i,j)(t) \\ \lambda(3)p(3,1) & \lambda(3)p(3,2) & -\lambda(3) & \cdots & P_{3n}(t) \\ \vdots & \vdots & \vdots & \cdots & \vdots \\ \lambda(n)p(n,1) & \lambda(n)p(n,2) & \lambda(n)p(n,3) & \cdots & -\lambda(n) \end{pmatrix}$$

$$\times \begin{pmatrix} P_{11}(t) & P_{12}(t) & P_{13}(t) & \ldots & P_{1n}(t) \\ P_{21}(t) & P_{22}(t) & P_{23}(t) & \ldots & P_{2n}(t) \\ P_{31}(t) & P_{32}(t) & P_{33}(t) & \ldots & P_{3n}(t) \\ \vdots & \vdots & \vdots & \ldots & \vdots \\ P_{n1}(t) & P_{n2}(t) & P_{n3}(t) & \ldots & P_{nn}(t) \end{pmatrix}$$

and

$$(GP_t)_{ij} = \sum_k G_{ik}(P_t)_{kj} = \sum_{k:k\neq i} G_{ik}(P_t)_{kj} + G_{ii}(P_t)_{ij} = \sum_{k:k\neq i}\lambda(i)p(i,k)(P_t)_{kj} - \lambda(i)(P_t)_{ij}.$$

From equation (6), we have

$$P'_{ij}(t) = -\lambda(i)P_{ij}(t) + \sum_{k:k\neq i}\left[p(i,k)\lambda(j)\right]P_{kj}(t) = (GP_t)_{ij}$$

so

$$\frac{d}{dt}(P_t) = GP_t. \tag{7}$$

The solution for equation (7) is

$$P_t = P_0 e^{tG}.$$

Thus, the solutions for the forward and backward equations appear the same. However, we will see later that while the backward equations are always valid, this is not the case for the forward equations.

Example

Consider a two-state process. This could be a model for an on–off system or a working-under repair system. We denote states {0,1} and suppose that state 0 changes at exponential rate α and state 1 changes at exponential rate β.

We will use the matrix

$$P(t) = \left(P_{ij}(t)\right)$$

where

$$P_{ij}(t) = P\left(X_t = j \middle| X_0 = i\right).$$

In this example,

$$G = \begin{pmatrix} -\alpha & \alpha \\ \beta & -\beta \end{pmatrix} \quad \text{and} \quad P(t) = \begin{pmatrix} P_{00}(t) & P_{01}(t) \\ P_{10}(t) & P_{11}(t) \end{pmatrix}.$$

The backward equation is

$$P'(t) = \begin{pmatrix} P'_{00}(t) & P'_{01}(t) \\ P'_{10}(t) & P'_{11}(t) \end{pmatrix} = GP(t) = \begin{pmatrix} -\alpha & \alpha \\ \beta & -\beta \end{pmatrix}\begin{pmatrix} P_{00}(t) & P_{01}(t) \\ P_{10}(t) & P_{11}(t) \end{pmatrix}$$

$$= \begin{pmatrix} -\alpha P_{00}(t) + \alpha P_{10}(t) & -\alpha P_{01}(t) + \alpha P_{11}(t) \\ \beta P_{00}(t) - \beta P_{10}(t) & \beta P_{01}(t) - \beta P_{11}(t) \end{pmatrix}.$$

and we have the initial conditions

$$P_{00}(0) = P_{11}(0) = 1, \quad P_{10}(0) = P_{01}(0) = 0.$$

In words, the states do not change at time 0.

This gives four ODEs:

$$P'_{00}(t) = -\alpha P_{00}(t) + \alpha P_{10}(t), \quad P_{00}(0) = 1. \tag{a}$$

$$P'_{01}(t) = -\alpha P_{01}(t) + \alpha P_{11}(t), \quad P_{01}(0) = 0. \tag{b}$$

$$P'_{10}(t) = \beta P_{00}(t) - \beta P_{10}(t), \quad P_{10}(0) = 0. \tag{c}$$

$$P'_{11}(t) = \beta P_{01}(t) - \beta P_{11}(t), \quad P_{11}(0) = 1. \tag{d}$$

Consider equations (a) and (c):

$$P'_{00}(t) = -\alpha P_{00}(t) + \alpha P_{10}(t) \quad \text{or} \quad \beta P'_{00}(t) = \beta\left[-\alpha P_{00}(t) + \alpha P_{10}(t)\right]$$

$$P'_{10}(t) = \beta P_{00}(t) - \beta P_{10}(t) \quad \text{or} \quad \alpha P'_{10}(t) = \alpha\left[\beta P_{00}(t) - \beta P_{10}(t)\right].$$

So

$$\beta P'_{00}(t) + \alpha P'_{10}(t) = 0$$

and integrating gives

$$\beta P_{00}(t) + \alpha P_{10}(t) = C.$$

The initial conditions $P_{00}(0) = 1$ and $P_{10}(0) = 0$ give $C = \beta$ so

$$\beta P_{00}(t) + \alpha P_{10}(t) = \beta$$

and thus

$$\alpha P_{10}(t) = \beta\left[1 - P_{00}(t)\right].$$

So

$$P'_{00}(t) = -\alpha P_{00}(t) + \alpha P_{10}(t) = -\alpha P_{00}(t) + \beta\left[1 - P_{00}(t)\right]$$

and

$$P'_{00}(t) + (\alpha + \beta) P_{00}(t) = \beta. \tag{8}$$

The solution to the homogeneous equation

$$P'_{00}(t) + (\alpha + \beta) P_{00}(t) = 0$$

or

$$\frac{P'_{00}(t)}{P_{00}(t)} = -(\alpha + \beta)$$

is determined by

$$\int \frac{P'_{00}(t)}{P_{00}(t)} dt = -(\alpha + \beta) \int dt.$$

Thus,

$$\ln P_{00}(t) = -(\alpha + \beta)t + C$$

and

$$P_{00}(t) = Ke^{-(\alpha+\beta)t} \quad \text{where } K = e^{C}.$$

By inspection, a particular solution to

$$P'_{00}(t) + (\alpha + \beta) P_{00}(t) = \beta$$

is

$$P_{00}(t) = \frac{\beta}{\alpha + \beta}$$

so the general solution to equation (8) is

$$P_{00}(t) = \frac{\beta}{\alpha + \beta} + Ke^{-(\alpha+\beta)t}.$$

Since $P_{00}(0) = 1$, we have

$$1 = \frac{\beta}{\alpha + \beta} + K$$

so

$$K = \frac{\alpha}{\alpha + \beta}$$

and thus

$$P_{00}(t) = \frac{\beta}{\alpha + \beta} + \frac{\alpha}{\alpha + \beta} e^{-(\alpha+\beta)t}.$$

Since $P_{00}(t) + P_{01}(t) = 1$, we conclude

$$P_{01}(t) = 1 - P_{00}(t) = \frac{\alpha + \beta}{\alpha + \beta} - \left[\frac{\beta}{\alpha + \beta} + \frac{\alpha}{\alpha + \beta} e^{-(\alpha+\beta)t} \right]$$

$$= \frac{\alpha}{\alpha + \beta} - \frac{\alpha}{\alpha + \beta} e^{-(\alpha+\beta)t}.$$

Also, by symmetry,

$$P_{11}(t) = \frac{\alpha}{\alpha + \beta} + \frac{\beta}{\alpha + \beta} e^{-(\alpha+\beta)t}$$

and

$$P_{10}(t) = 1 - P_{11}(t) = \frac{\beta}{\alpha + \beta} - \frac{\beta}{\alpha + \beta} e^{-(\alpha+\beta)t}.$$

The long-term behavior can be found by taking the limit as $t \to \infty$.

Suppose the system starts in state 0 and we want to know the proportion of time the system is in state 1 for a very long period of time. Then we compute

$$\lim_{t \to \infty} P_{01}(t) = \lim_{t \to \infty} \left[\frac{\alpha}{\alpha + \beta} - \frac{\alpha}{\alpha + \beta} e^{-(\alpha+\beta)t} \right] = \frac{\alpha}{\alpha + \beta}.$$

Likewise, if the system starts in state 1 and we want to know the proportion of time the system is in state 1 for a very long period of time, then

$$\lim_{t \to \infty} P_{11}(t) = \lim_{t \to \infty} \left[\frac{\alpha}{\alpha + \beta} + \frac{\alpha}{\alpha + \beta} e^{-(\alpha+\beta)t} \right] = \frac{\alpha}{\alpha + \beta}.$$

It should not be surprising that the initial state of the system does not affect the long-term behavior.

We repeat this problem for the forward equation

$$P'(t) = P(t)G.$$

Now

$$P'(t) = \begin{pmatrix} P'_{00}(t) & P'_{01}(t) \\ P'_{10}(t) & P'_{11}(t) \end{pmatrix} = P(t)G = \begin{pmatrix} P_{00}(t) & P_{01}(t) \\ P_{10}(t) & P_{11}(t) \end{pmatrix} \begin{pmatrix} -\alpha & \alpha \\ \beta & -\beta \end{pmatrix}$$

$$= \begin{pmatrix} -P_{00}(t)\alpha + P_{01}(t)\beta & P_{00}(t)\alpha - P_{01}(t)\beta \\ -P_{10}(t)\alpha + P_{11}(t)\beta & P_{10}(t)\alpha - P_{11}(t)\beta \end{pmatrix}$$

so we have the four ODEs:

$$P'_{00}(t) = -P_{00}(t)\alpha + P_{01}(t)\beta.$$

$$P'_{01}(t) = P_{00}(t)\alpha - P_{01}(t)\beta.$$

$$P'_{10}(t) = -P_{10}(t)\alpha + P_{11}(t)\beta.$$

$$P'_{11}(t) = P_{10}(t)\alpha - P_{11}(t)\beta.$$

Consider the first equation

$$P'_{00}(t) = -P_{00}(t)\alpha + P_{01}(t)\beta = -P_{00}(t)\alpha + \left[1 - P_{00}(t)\right]\beta$$

or

$$P'_{00}(t) + (\alpha + \beta)P_{00}(t) = \beta,$$

which is the same equation as in the backward Kolmogorov equation. We leave it as an exercise to show the remaining three equations are the same.

Example

In a power production system, there are a series of generators. The time until failure of each generator is exponentially distributed with mean 100 h and the time of repair is exponentially distributed with a mean of 40 h.

(a) Suppose there is one generator and at time $t = 0$ it is operational. What is the probability that at $t = 150$ the generator is operational?

Solution

Let state 0 denote the generator is operating and state 1 denote the generator is down. The rate for operating is

$$\alpha = \frac{1}{100} = .01$$

and the rate of repair is

$$\beta = \frac{1}{40} = .025.$$

We compute

$$P_{00}(t) = \frac{\beta}{\alpha + \beta} + \frac{\alpha}{\alpha + \beta} e^{-(\alpha+\beta)t}$$

for $t = 150,\ \alpha = .01, \text{and}\ \beta = .025$ to get

$$P_{00}(150) = \frac{.025}{.01 + .025} + \frac{.01}{.01 + .025} e^{-(.01+.025)150} \approx .716.$$

(b) Suppose there are 3 identical generators that operate independently, and the system is up if at least 1 of the generators is up. What is the probability the system is up after 150 h?

Solution

For the system to be down, all 3 must fail. The probability of all three failing at 150 h is

$$(1 - .716)^3 = .023$$

so the probability that at least one of the generators is working is

$$1 - .023 = .977.$$

Connections of the Infinitesimal Generator, the Embedded Markov Chain, Transition Rates, and the Stationary Distribution

We demonstrate that there is a one-to-one association between continuous-time Markov chains and square matrices G that satisfy the conditions of equation (2).

Example

Suppose that G is a matrix satisfying the conditions of (2), say

$$G = \begin{pmatrix} -10 & 3 & 2 & 5 \\ 0 & -4 & 1 & 3 \\ 2 & 1 & -5 & 2 \\ 0 & 2 & 0 & -2 \end{pmatrix}.$$

We want to recover the rates of change and the embedded Markov chain associated with G. Recall that for $i \neq j$, we have

$$G_{ij} = \lambda(i)p(i,j) \quad \text{so } p(i,j) = \frac{G_{ij}}{\lambda(i)} = \frac{\lambda(i)p(i,j)}{\lambda(i)}$$

and

$$\lambda(i) = \sum_{j:j \neq i} G_{ij} = \sum_{j:j \neq i} \lambda(i)p(i,j) = -G_{ii}.$$

Thus, we can recover $\lambda(i)$ and $p(i, j)$ from G. Recall that $p(i, j)$ are the entries of the embedded Markov chain.

In our example,

$$\lambda(1) = 3+2+5 = 10; \quad p(1,1) = 0,\ p(1,2) = \frac{3}{10},\ p(1,3) = \frac{2}{10},\ p(1,4) = \frac{5}{10}$$

$$\lambda(2) = 0+1+3 = 4; \quad p(2,1) = \frac{0}{4},\ p(2,2) = 0,\ p(2,3) = \frac{1}{4},\ p(2,4) = \frac{3}{4}$$

$$\lambda(3) = 2+1+2 = 5; \quad p(3,1) = \frac{2}{5},\ p(3,2) = \frac{1}{5},\ p(3,3) = 0,\ p(3,4) = \frac{2}{5}$$

$$\lambda(4) = 0+2+0 = 2; \quad p(4,1) = \frac{0}{2},\ p(4,2) = \frac{2}{2},\ p(4,3) = \frac{0}{2},\ p(4,4) = 0.$$

So the embedded Markov chain has the transition matrix

$$\begin{pmatrix} 0 & .3 & .2 & .5 \\ 0 & 0 & .25 & .75 \\ .4 & .2 & 0 & .4 \\ 0 & 1 & 0 & 0 \end{pmatrix}.$$

Note that the diagonal elements are 0.

On the other hand, if one knows the change rates and the embedded Markov chain, then

$$G_{ij} = \lambda(i)p(i,j) \quad \text{if } i \neq j \text{ and } G_{ii} = -\lambda(i).$$

Note that if P is a stochastic matrix, then $P - I$ satisfies these conditions.

Theorem 5.2

Suppose that $\{X_t\}$ is an irreducible finite-state continuous-time Markov process with infinitesimal generator G. Then

(a) G has an eigenvalue of 0. The associated eigenvector has algebraic and geometric dimension 1. If $\hat{\pi}$ is such an eigenvector, then $\hat{\pi}$ can be chosen so that it is a probability vector with all components positive. Every other eigenvalue of G has negative real part.
(b) For any probability vector $\hat{v}$, we have

$$\lim_{t\to\infty} \hat{v} e^{tG} = \hat{\pi}$$

where $\hat{\pi}$ is the unique probability vector for which $\hat{\pi}G = \hat{0}$.

Proof

(a) The proof will involve manipulations that enable us to apply the Perron–Frobenius theorem.

Let K be a number larger than $\max|G_{ij}|$ (this is one of the places where finiteness is needed; another is the Perron–Frobenius theorem) and let

$$G^* = \frac{1}{K}G.$$

Since $G_{ij} \geq 0$ if $i \neq j$, and $\sum_j G_{ii} = 0$, then

$$G^*_{ij} \geq 0 \quad \text{if } i \neq j \text{ and } \sum_j G^*_{ij} = 0 \text{ for all } i.$$

Let

$$P = G^* + I.$$

Then

$$P_{ij} = G^*_{ij} \geq 0 \quad \text{if } i \neq j \text{ and } P_{ii} = G^*_{ii} + 1 = -\sum_{j:j\neq i} G^*_{ij} + 1$$

and thus

$$\sum_j P_{ij} = \sum_{j:j\neq i} G^*_{ij} + \left(-\sum_{j:j\neq i} G^*_{ij} + 1 \right) = 1.$$

So P has the properties $P_{ij} \geq 0$ and $\sum_j P_{ij} = 1$ for all i; thus, P is a stochastic matrix. Also, $\hat{v}$ is an eigenvector for P if and only if $\hat{v}$ is an eigenvector for G and G^* for if $\hat{v}G = \lambda\hat{v}$, then

$$\hat{v}G^* = \hat{v}\frac{1}{K}G = \frac{1}{K}\hat{v}G = \frac{\lambda}{K}\hat{v}$$

and

$$\hat{v}P = \hat{v}\left(G^* + I\right) = \left(\frac{\lambda}{K} + 1\right)\hat{v}.$$

Since P is a stochastic matrix for an irreducible Markov chain, the Perron–Frobenius theorem applies, and 1 is an eigenvalue of P with algebraic and geometric dimension one, and the eigenvector for the eigenvalue 1 can be chosen to be a probability vector with positive entries. Furthermore, if λ is any other eigenvalue of P, then $|\lambda| < 1$ so the real part of $\lambda < 1$.

Since $G^* = P - I$, if λ is an eigenvalue of P, then $\lambda - 1$ is an eigenvalue of G^* so 0 is an eigenvalue of G^* and any other eigenvalue of G^* has a negative real part. Thus, 0 is an eigenvalue of G and any other eigenvalue of G has a negative real part.

(b) To determine the equilibrium state, we follow the same ideas as in the discrete case. Let G be the infinitesimal generator of an irreducible Markov chain. Then G can be diagonalized if and only if there is a basis of eigenvectors. If that is the case, then there is a matrix Q and a diagonal matrix D for which

$$D = Q^{-1}GQ \quad \text{or} \quad G = QDQ^{-1}.$$

Recall from linear algebra that the columns of Q are the eigenvectors of G and the entries on the diagonal of D are the eigenvalues of G.

We have

$$P_t = e^{tG} = \sum_{n=0}^{\infty} \frac{(tG)^n}{n!} = \sum_{n=0}^{\infty} \frac{t^n \left(QDQ^{-1}\right)^n}{n!} = Q\left[\sum_{n=0}^{\infty} \frac{t^n D^n}{n!}\right]Q^{-1}.$$

Now if

$$D = \begin{pmatrix} d_1 & 0 & \dots & 0 \\ 0 & d_2 & \dots & 0 \\ \vdots & \vdots & \vdots & \vdots \\ 0 & 0 & \dots & d_k \end{pmatrix},$$

then

$$D^n = \begin{pmatrix} d_1{}^n & 0 & \dots & 0 \\ 0 & d_2{}^n & \dots & 0 \\ \vdots & \vdots & \vdots & \vdots \\ 0 & 0 & \dots & d_k{}^n \end{pmatrix}$$

so

$$Q\left[\sum_{n=0}^{\infty} \frac{t^n D^n}{n!}\right]Q^{-1} = Q\sum_{n=0}^{\infty} \begin{pmatrix} (td_1)^n/n! & 0 & \dots & 0 \\ 0 & (td_2)^n/n! & \dots & 0 \\ \vdots & \vdots & \vdots & \vdots \\ 0 & 0 & \dots & (td_k)^n/n! \end{pmatrix} Q^{-1}$$

$$= Q\begin{pmatrix} e^{td_1} & 0 & \dots & 0 \\ 0 & e^{td_2} & \dots & 0 \\ \vdots & \vdots & \vdots & \vdots \\ 0 & 0 & \dots & e^{td_k} \end{pmatrix} Q^{-1}. \tag{9}$$

We can arrange the eigenvalues so that $d_1 = 0$ and then every other d_i has negative real part $-r_i$, so that

$$e^{tG} = Q\begin{pmatrix} e^{td_1} & 0 & \dots & 0 \\ 0 & e^{td_2} & \dots & 0 \\ \vdots & \vdots & \vdots & \vdots \\ 0 & 0 & \dots & e^{td_k} \end{pmatrix}Q^{-1} = Q\begin{pmatrix} e^{0} & 0 & \dots & 0 \\ 0 & \alpha_2 e^{-tr_2} & \dots & 0 \\ \vdots & \vdots & \vdots & \vdots \\ 0 & 0 & \dots & \alpha_k e^{-tr_k} \end{pmatrix}Q^{-1}$$

$$= Q\begin{pmatrix} 1 & 0 & \dots & 0 \\ 0 & 0 & \dots & 0 \\ \vdots & \vdots & \vdots & \vdots \\ 0 & 0 & \dots & 0 \end{pmatrix}Q^{-1} + \alpha_2 e^{-tr_2} Q\begin{pmatrix} 0 & 0 & \dots & 0 \\ 0 & 1 & \dots & 0 \\ \vdots & \vdots & \vdots & \vdots \\ 0 & 0 & \dots & 0 \end{pmatrix}Q^{-1} + \cdots$$

$$+\alpha_k e^{-tr_k} Q\begin{pmatrix} 0 & 0 & \dots & 0 \\ 0 & 0 & \dots & 0 \\ \vdots & \vdots & \vdots & \vdots \\ 0 & 0 & \dots & 1 \end{pmatrix}Q^{-1}$$

where $\alpha_j = e^{\mathrm{Im}(td_j)}$, so $|\alpha_j| = 1$.

Thus, since

$$\lim_{t\to\infty} e^{-tr_i} = 0,$$

we have

$$\lim_{t\to\infty} P_t = Q\begin{pmatrix} 1 & 0 & \dots & 0 \\ 0 & 0 & \dots & 0 \\ \vdots & \vdots & \vdots & \vdots \\ 0 & 0 & \dots & 0 \end{pmatrix}Q^{-1} = \begin{pmatrix} \hat{\pi} \\ \hat{\pi} \\ \vdots \\ \hat{\pi} \end{pmatrix}$$

where the last expression is the matrix with every row the probability eigenvector whose eigenvalue is 0.

If it is not the case that there is a basis of eigenvectors, then it can be shown that the result is still valid using the methods of Chapter 2.

Example

Consider

$$G = \begin{pmatrix} -3 & 1 & 2 \\ 4 & -5 & 1 \\ 1 & 1 & -2 \end{pmatrix}.$$

The eigenvalues of G together with their respective eigenvectors are

$$\lambda_1 = 0,\ \hat{v}_1 = \begin{pmatrix} 1 \\ 1 \\ 1 \end{pmatrix};\quad \lambda_2 = -4,\ \hat{v}_2 = \begin{pmatrix} -3 \\ -7 \\ 5 \end{pmatrix};\quad \lambda_3 = -6,\ \hat{v}_3 = \begin{pmatrix} 1 \\ -5 \\ 1 \end{pmatrix}.$$

Then

$$Q = \begin{pmatrix} 1 & -3 & 1 \\ 1 & -7 & -5 \\ 1 & 5 & 1 \end{pmatrix},\quad Q^{-1} = \begin{pmatrix} \frac{3}{8} & \frac{1}{6} & \frac{11}{24} \\ -\frac{1}{8} & 0 & \frac{1}{8} \\ \frac{1}{4} & -\frac{1}{6} & -\frac{1}{12} \end{pmatrix}$$

$$D = \begin{pmatrix} 0 & 0 & 0 \\ 0 & -4 & 0 \\ 0 & 0 & -6 \end{pmatrix}$$

and

$$\begin{aligned} P_t &= Qe^{tD}Q^{-1} \\ &= \begin{pmatrix} \frac{3}{8} & \frac{1}{6} & \frac{11}{24} \\ \frac{3}{8} & \frac{1}{6} & \frac{11}{24} \\ \frac{3}{8} & \frac{1}{6} & \frac{11}{24} \end{pmatrix} + e^{-4t}\begin{pmatrix} \frac{3}{8} & 0 & -\frac{3}{8} \\ \frac{7}{8} & 0 & -\frac{7}{8} \\ -\frac{5}{8} & 0 & \frac{5}{8} \end{pmatrix} + e^{-6t}\begin{pmatrix} \frac{1}{4} & -\frac{1}{6} & -\frac{1}{12} \\ -\frac{5}{4} & \frac{5}{6} & \frac{5}{12} \\ \frac{1}{4} & -\frac{1}{6} & -\frac{1}{12} \end{pmatrix}. \end{aligned}$$

So the equilibrium distribution is $\hat{\pi} = \begin{pmatrix} \frac{3}{8} & \frac{1}{6} & \frac{11}{24} \end{pmatrix}$.

We can also solve for $(P_t)_{ij}$. We have

$$P_t = Qe^{tD}Q^{-1} = \begin{pmatrix} 1 & -3 & 1 \\ 1 & -7 & -5 \\ 1 & 5 & 1 \end{pmatrix} \begin{pmatrix} 1 & 0 & 0 \\ 0 & e^{-4t} & 0 \\ 0 & 0 & e^{-6t} \end{pmatrix} \begin{pmatrix} 1 & -3 & 1 \\ 1 & -7 & -5 \\ 1 & 5 & 1 \end{pmatrix}^{-1}$$

$$= \begin{pmatrix} \frac{3e^{-4t}}{8} + \frac{e^{-6t}}{4} + \frac{3}{8} & \frac{1}{6} - \frac{e^{-6t}}{6} & \frac{-3e^{-4t}}{8} - \frac{e^{-6t}}{12} + \frac{11}{24} \\ \frac{7e^{-4t}}{8} - \frac{5e^{-6t}}{4} + \frac{3}{8} & \frac{5e^{-6t}}{4} + \frac{1}{6} & \frac{5e^{-6t}}{6} - \frac{7e^{-4t}}{8} + \frac{11}{24} \\ -\frac{5e^{-4t}}{8} + \frac{e^{-6t}}{4} + \frac{3}{8} & \frac{1}{6} - \frac{e^{-6t}}{6} & \frac{5e^{-4t}}{8} - \frac{e^{-6t}}{12} + \frac{11}{24} \end{pmatrix}.$$

So we know $(P_t)_{ij}$. For example, the probability that the process begins in state 2 and is in state 3 at time t is

$$\frac{5e^{-6t}}{6} - \frac{7e^{-4t}}{8} + \frac{11}{24}$$

Connection between the Steady State of a Continuous Markov Chain and the Steady State of the Embedded Matrix

Theorem 5.3

Let $X(t)$ be an irreducible, positive recurrent continuous-time Markov chain with infinitesimal generator G. Let P be the transition matrix of the embedded Markov chain. Let π be the unique equilibrium state of the embedded Markov chain and $\lambda_k = -G_{kk}$. Then the vector φ defined by

$$\varphi_k = \frac{\pi_k}{\lambda_k}$$

is invariant for $X(t)$. If $\sum \varphi_k < \infty$, then the vector φ can be normalized, and the normalized vector is the unique equilibrium state for $X(t)$.

Proof

We have π is the unique probability distribution for which $\pi = \pi P$. Also

$$G_{ij} = \begin{cases} \lambda_i P(i,j) & \text{if } i \neq j \\ -\lambda_i & \text{if } i = j \end{cases}.$$

Now

$$\varphi G = \hat{0} \quad \text{if and only if} \sum_j \varphi_i G_{ij} = 0 \quad \text{for all } j$$

and

$$\sum_j \varphi_i G_{ij} = \sum_{j:j\neq i} \varphi_i G_{ij} + \varphi_i G_{ii} = \sum_{j:j\neq i} \varphi_i \lambda_i P(i,j) - \varphi_i \lambda_i.$$

So

$$\varphi G = \hat{0} \quad \text{if and only if} \sum_{j:j\neq i} \varphi_i \lambda_i P(i,j) = \varphi_i \lambda_i,$$

which is true if and only if $\pi = \pi P$ with $\pi_i = \varphi_i \lambda_i$.

Example

Suppose that $X(t)$ is a continuous-time Markov chain with infinitesimal generator

$$G = \begin{pmatrix} -3 & 1 & 2 \\ 4 & -5 & 1 \\ 1 & 1 & -2 \end{pmatrix}.$$

Recall that the transition matrix P for the embedded Markov chain is given by

$$P_{ij} = \begin{cases} \dfrac{G_{ij}}{G_i} & \text{if } i \neq j \quad \text{where } G_i = \sum_{j:j\neq i} G_{ij} \\ 0 & \text{if } i = j \end{cases}$$

so $G_1 = 3$, $G_2 = 5$, $G_3 = 2$ and

$$P = \begin{pmatrix} 0 & \frac{1}{3} & \frac{2}{3} \\ \frac{4}{5} & 0 & \frac{1}{5} \\ \frac{1}{2} & \frac{1}{2} & 0 \end{pmatrix}.$$

An eigenvector $\hat{x}$ satisfying $\hat{x}P = \hat{x}$ is (27,20,22), which can be normalized to give

$$\pi = \left(\frac{27}{69}, \frac{20}{69}, \frac{22}{69}\right).$$

A vector τ satisfying $\tau G = \hat{0}$ is τ = (9,4,11), which can be normalized to give

$$\varphi = \left(\frac{9}{24}, \frac{4}{24}, \frac{11}{24}\right).$$

So

$$(\varphi_1\lambda_1, \varphi_2\lambda_2, \varphi_3\lambda_3) = \left(\frac{9}{24}\cdot 3, \frac{4}{24}\cdot 5, \frac{11}{24}\cdot 2\right) = \left(\frac{27}{24}, \frac{20}{24}, \frac{22}{24}\right) = \frac{69}{24}\left(\frac{27}{69}, \frac{20}{69}, \frac{22}{69}\right)$$

$$= \frac{69}{24}\pi.$$

In studying discrete Markov chains, the simplest case was when the chain was irreducible and aperiodic. The reducible chains were seen to behave like a group of irreducible chains each acting on a single class, pasted together. We do the same thing for continuous Markov chains. The next theorem shows there are no periodic states in an irreducible continuous Markov chain.

Theorem 5.4

If X_t is an irreducible continuous Markov chain, then for each pair of states i and j, and each $t > 0$,

$$P(X_t = j \mid X_0 = i) > 0.$$

Proof

Fix states i and j and $t > 0$. Since X_t is irreducible, there are states $s_1, \ldots, s_n$ for which $p(i, s_1)$, $p(s_1, s_2)$, $p(s_2, s_3)$, …, $p(s_n, j)$ are all positive. Thus, it is possible to go from state i to state j in $n + 1$ steps. Let $\lambda_0, \lambda_1, \ldots, \lambda_n$ be the respective transition rates from states $i, s_1, \ldots, s_n$. Then

$$\left[1-e^{-(\lambda_0 t_0/(n+1))}\right]p(i,s_1)\left[1-e^{-(\lambda_1 t_0/(n+1))}\right]p(s_1,s_2)\cdots\left[1-e^{-\left(\frac{\lambda_n t_0}{(n+1)}\right)}\right]p(s_n,j) > 0$$

so

$$P(X_{t_0} = j \mid X_0 = i) > 0.$$

Explosions

In a continuous infinite-state Markov chain, it is possible that an infinite number of transitions could occur in a finite amount of time. When that happens, we say that an explosion occurs. (We give a formal definition in the following.) In this section, we determine conditions on the exponential holding times that will cause an explosion to occur.

Let X_t be a continuous Markov process, and let $\lambda(i)$ be the exponential change rate when the system is in state i. This means the expected length of time in state i is $1/\lambda(i)$. It should seem reasonable that the larger the $\lambda(i)$, the shorter the stay in state i will be and that more transitions will occur. This, in turn, makes an explosion more likely.

Example

Consider a continuous Markov process where the state space is {1,2,3,…} and $P(n,n+1) = 1$. Suppose that the exponential change rate when the system is in state n is n^2. Then the expected amount of time for the first N changes is

$$\sum_{n=1}^{N} \frac{1}{n^2}.$$

Since this series converges, it would seem that it is possible for there to be infinitely many changes in a finite amount of time. We will prove this is the case in Theorem 5.5.

Let T_n be the time of the nth transition of the process. We define

$$T_\infty = \lim_{n\to\infty} T_n.$$

Note that T_∞ can be finite or infinite. If T_∞ is finite, then there were infinitely many transitions in a finite amount of time; that is, an explosion occurred.

Definition

Let i be a state in a continuous Markov process $X(t)$ and let

$$P_i(T_\infty = \infty) = P\big(T_\infty = \infty \mid X(0) = i\big).$$

If $P_i(T_\infty = \infty) = 1$ for every state i, then the process is said to be nonexplosive. Otherwise, the process is said to be explosive.

The following example gives a Markov process where there is state i for which $P_i(T_\infty = \infty) = 1$ and state j for which $P_j(T_\infty = \infty) = 0$. According to the definition, this process is explosive.

Example

Consider the continuous infinite-state Markov chain whose digraph is shown in Figure 5.1.

If the process is in state −2, it goes to state −1 with probability 1, and if the process is in state −1, it goes to state −2 with probability 1.

If the process is in state 0, it goes to state −1 w with probability 1/2 and goes to state 1 with probability 1/2.

For state $n \geq 1$, the process goes to state $n + 1$ with probability 1.

Suppose further that the transition rate from states −2, −1, and 0 is 1 and the transition rate from state n is n^2 for $n \geq 1$. We show later that if the initial state is $n \geq 1$, then an explosion occurs.

In our study of the Kolmogorov forward and backward equations, we noted that the backward equations were always valid, but this is not the case with the forward equations. The forward equations are valid if and only if there are no explosions. There are various heuristic descriptions for why this is the case. Robert Feller (1968, p. 451) describes the difficulty as that for

$$P_i(t) = e^{-\lambda_i t}$$

there needs to be a proper probability distribution, which means

$$\sum P_i(t) = 1$$

for all t, whereas with an explosion

$$\sum P_i(t) < 1$$

for some values of t. A more technical explanation is that the infinitesimal generator can be an unbounded operator, in which case problems with the domain can arise. Another problem is what to do at T_∞. (One approach would require the process to be absorbed at T_∞. Another is to immediately pass to a state k at T_∞ with probability q_k.)

The most intuitive examples of Markov chains that can be explosive are pure birth processes, where the state space is the positive integers. For pure birth processes, the process can only transition from n to $n + 1$ and the holding time in state n is exponentially distributed with transition rate $\lambda(n)$. If $\lambda(n) = n^2$, for example, then the process would be explosive as the next theorem shows.

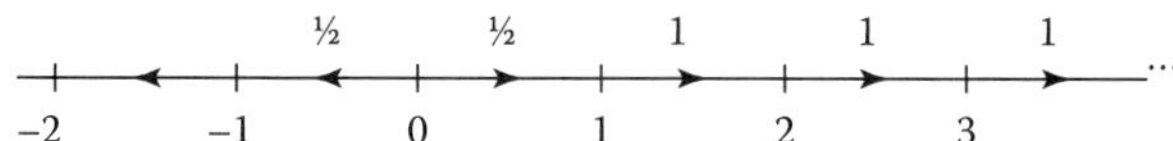

FIGURE 5.1
Digraph for Markov chain.

Theorem 5.4

Let $\{S_n\}$ be an infinite collection of random variables that are independent and exponentially distributed. Let $\lambda(n)$ be the parameter of S_n.

Then $\sum_n S_n < \infty$ with probability 1 if and only if $\sum_n (1/\lambda(n))$ converges.

Proof

Note that

$$E\left[\sum_{n=1}^{\infty} S_n\right] = \sum_{n=1}^{\infty} E[S_n] = \sum_{n=1}^{\infty} \frac{1}{\lambda(n)},$$

and if

$$E\left[\sum_{n=1}^{\infty} S_n\right] < \infty,$$

then

$$P\left(\sum_{n=1}^{\infty} S_n = \infty\right) = 0.$$

To prove the converse, we use the fact that

$$\prod (1+a_n)^{-1}$$

converges to 0 if and only if $\sum a_n$ diverges to ∞.

Sketch of a proof:

$$\ln\left[\prod (1+a_n)^{-1}\right] = -\sum \ln(1+a_n).$$

Also

$$\lim_{x\to 0} \frac{\ln(1+x)}{x} = 1$$

So if $a_n \to 0$, by the limit comparison test,

$$\prod (1+a_n)^{-1}$$

converges to 0 if and only if $\sum a_n$ diverges to ∞.

Suppose that the random variable T is exponentially distributed with parameter λ. Then

$$E\left[e^{-T}\right]=\int_0^\infty e^{-t}\lambda e^{\lambda t}dt=\frac{\lambda}{1+\lambda}\int_0^\infty e^{-(1+\lambda)t}dt=\frac{\lambda}{1+\lambda}=\frac{1}{1+\frac{1}{\lambda}}.$$

So if $\{T_n\}$ is a collection of independent exponentially distributed random variables, and the parameter of T_n is λ_n, then

$$E\left[e^{-(T_1+T_2+\cdots)}\right]=\prod\left(1+\frac{1}{\lambda_n}\right)^{-1}.$$

If

$$\sum\frac{1}{\lambda_n}$$

diverges to ∞, then

$$=\prod\left(1+\frac{1}{\lambda_n}\right)^{-1}=0.$$

But

$$e^{-(T_1+T_2+\cdots)}$$

is a nonnegative function, and its expectation is *0* if and only if the function is 0 with probability 1, that is, if $\sum T_n=\infty$ with probability 1.

Corollary

A pure birth process with transition rates $\lambda(n)$ is explosive if and only if

$$\sum_{n=1}^{\infty}\frac{1}{\lambda(n)}<\infty.$$

Proof

Let $T_0 = 0$, and for $n > 0$, let T_n be the time between the $(n - 1)$st birth and the nth birth. Then T_n is exponentially distributed and

$$E[T_n] = \frac{1}{\lambda(n)}.$$

Then $S_N = T_1 + \cdots + T_N$ is the time of the nth birth and

$$E[S_N] = E[T_1 + \cdots + T_N] = \sum_{k=1}^{N} E[T_k] = \sum_{k=1}^{N} \frac{1}{\lambda(k)},$$

and so $E[S_N] < \infty$ if and only if

$$\sum_{k=1}^{\infty} \frac{1}{\lambda(k)}$$

converges.

Birth and Birth–Death Processes

Among the simplest continuous-time Markov processes are pure birth processes where members can only be added to a population (through births) and birth–death processes where the population can increase or decrease. These are also among the most important examples because of the number of situations that they describe. In this section, we develop the Kolmogorov equations for these processes and we use them to describe examples of the forward and backward equations.

In general, the forward and backward equations describe the dynamics of a diffusion process. The birth and birth–death processes are simple examples of a diffusion process.

For the forward equation, we know the state of the process at time t and want to predict the state of the process at some later time. The later time will be $t + h$ and we will use the postulates of the system to arrive at a differential equation.

In the backward equation, we know the state of the process at time t and want to determine the probability it was in various states at a previous time. We again use the postulates to arrive at a differential equation. Many sources describe this as *integrating backward.*

Birth Process (Forward Equation)

A birth process (also called a pure birth process) is a continuous Markov chain where the state space is the set of positive integers and the only transitions possible are $k \to k+1$.

To give the Kolmogorov equations for the birth and birth–death processes, we recall that the Kolmogorov forward equation is

$$P_{ij}' = (t) - \lambda(j)P_{ij}(t) + \sum_{k:k\neq j}\left[\lambda(k)p(k,j)\right]P_{ik}(t).$$

The notation in the literature for birth and birth–death processes with the Kolmogorov forward equation usually refers to these equations using $P_n(t)$ to denote the probability that the system is in state n at time t. This can be determined only if the distribution at time $t = 0$ is known. The usual way to address the discrepancy is to set $P_n(t) = P_{in}(t)$ with initial condition $P_i(0) = 1$ and $P_k(0) = 0$ if $k \neq i$. With this understanding, we will follow the notation for birth and birth–death processes that is common in the literature.

For the birth process, $p(n-1,n) = 1$ and $p(k, n) = 0$ if $k \neq n-1$. Thus, according to the general equation, the Kolmogorov forward equation for the birth process is

$$P_n'(t) = -\lambda(n)P_n(t) + \lambda(n-1)P_{n-1}(t).$$

We derive these equations from the first principles of a birth process.

In the birth process, we postulate that $\lambda_k \equiv \lambda_k(k)$ is the birth rate when the system is in state k, so that

(1) The probability of no jumps from state k in the interval of time $(t, t+h)$ is $1 - \lambda_k h + o(h)$
(2) The probability of exactly one jump from $k \to k+1$ in the time interval $(t, t+h)$ is $\lambda_k h + o(h)$
(3) The probability of more than one jump in the interval of time $(t, t+h)$ is $o(h)$

Let $P_n(t)$ denote the probability that the system is in state n at time t. We compute $P_n(t+h)$ where h is small. We examine how the system could be in state n at time $t+h$ based on the state of the system at time t and the associated probabilities.

1. The system could have been in state n at time t and made no transitions between t and $t+h$. The probability of no transitions between t and $t+h$ is $1 - \lambda_n h + o(h)$. Thus, the probability of being in state n at time t and remaining in that state until $t+h$ is $P_n(t)\,[1 - \lambda_n h + o(h)]$.
2. The system could have been in state $n-1$ at time t and made exactly one transition between t and $t+h$. The probability of exactly one

transition between t and $t + h$ is $\lambda_{n-1}h + o(h)$. Thus, the probability of being in state $n - 1$ at time t and being in state n at time $t + h$ as the result of exactly one transition is $P_{n-1}(t)\left[\lambda_{n-1}h + o(h)\right]$.

3. The system could be in state n at time $t + h$ as the result of more than one transition between t and $t + h$. The probability of this occurring is $o(h)$.

Thus,

$$\begin{aligned}
P_n(t+h) &= P_n(t)\left[1 - \lambda_n h + o(h)\right] + P_{n-1}(t)\left[\lambda_{n-1}h + o(h)\right] + o(h) \\
&= P_n(t)[1 - \lambda_n h] + P_{n-1}(t)[\lambda_n h] + o(h) \\
&= P_n(t) - P_n(t)\lambda_n h + P_{n-1}(t)\lambda_{n-1}h + o(h)
\end{aligned}$$

so that

$$\frac{P_n(t+h) - P_n(t)}{h} = -P_n(t)\lambda_n + P_{n-1}(t)\lambda_{n-1} + \frac{o(h)}{h}.$$

Then

$$P_n'(t) = \lim_{h\to 0}\frac{P_n(t+h) - P_n(t)}{h} = -\lambda_n P_n(t) + \lambda_{n-1}P_{n-1}(t)\ n \geq 1. \tag{9}$$

$$\text{For } n = 0,\ P_{n-1}(t) = 0 \text{ so } P_0'(t) = -\lambda_0 P_0(t). \tag{10}$$

Equations (9) and (10) are the forward equations and agree with what we observed earlier from the general theory.

From our postulates, we also get that the embedded Markov chain, E, and the infinitesimal generator, G, are given by

$$E = \begin{pmatrix} 0 & \lambda_0 & 0 & 0 & \cdots \\ 0 & 0 & \lambda_1 & 0 & \cdots \\ 0 & 0 & 0 & \lambda_2 & 0 \\ \vdots & \vdots & \vdots & \vdots & \vdots \end{pmatrix} \quad G = \begin{pmatrix} -\lambda_0 & \lambda_0 & 0 & 0 & \cdots \\ 0 & -\lambda_1 & \lambda_1 & 0 & \cdots \\ 0 & 0 & -\lambda_2 & \lambda_2 & 0 \\ \vdots & \vdots & \vdots & \vdots & \vdots \end{pmatrix}.$$

Earlier, we showed that the equilibrium state of a continuous Markov process with infinitesimal generator G was the unique probability vector $\hat{\pi}$ for which

$$\hat{\pi}G = \hat{0},$$

but here no such vector $\hat{\pi}$ exists. This should not be surprising because a birth process is one of unbounded growth.

Birth Process (Backward Equation)

The Kolmogorov backward equation for a general process is

$$P_{ij}'(t) = -\lambda(i)P_{ij}(t) + \sum_{k:k\neq i}\left[p(i,k)\lambda(j)h\right]P_{kj}(t).$$

In the backward equation, it is necessary to keep the $P_{ij}(t)$ notation (rather than using the $P_n(t)$ notation) and the backward equation for a birth process is

$$P_{in}'(t) = -\lambda_i P_{in}(t) + \lambda_i P_{i+1,n}(t).$$

We give detailed reasoning for this situation.

In the forward equations, we considered what changes could take place from t to time $t + h$ so that at time $t + h$ the system would be in state n. One way of looking at this is that we were lengthening the time interval from $(0, t)$ to $(0, t + h)$.

In the backward equations, we fix state n and consider how the system could have arrived at that state at the at time t by beginning at state i at time $-h$ for each of states i. Thus, we will have a different differential equation for each state i.

Here we use the notation

$P_{in}(t)$= the probability that given the system is in state i at time s, it will be in state n at times $s + t$. (This is true for any s by homogeneity, so it is not necessary for s to appear in $P_{in}(t)$.)

Fix state i. We consider how the system, starting in state i at time $-h$, can evolve to n at time t by breaking the interval $(-h, t)$ into $(-h, 0] \cup [0, t)$.

1. The system makes no transitions between $-h$ and 0. The probability of no transitions between $-h$ and 0 is $1 - \lambda_i h + o(h)$. The system must then jump to state n in the interval $(0, t)$. This occurs with probability $P_{in}(t)$. The probability of this sequence of events is $P_{in}(t)\left[1-\lambda_i h+o(h)\right]$.
2. In the interval $(-h, 0]$, the system makes exactly one transition. With the birth process, this must be to state $i + 1$. This occurs with probability $\lambda_i h + o(h)$. The system must then transition from state $i + 1$ to n in the interval $[0,t)$. This occurs with probability $P_{i+1,n}(t)$. The probability of this sequence of events is $\left[\lambda_i h+o(h)\right]P_{i+1,n}(t)$.
3. The system makes more than one transition in the interval $(-h, 0]$ and then transitions to state n in the interval $[0, t)$. This sequence of events occurs with probability $o(h)$.

Thus, we have

$$P_{in}(t+h) = P_{in}(t)\left[1-\lambda_i h+o(h)\right] + P_{i+1,n}(t)\left[\lambda_i h+o(h)\right] + o(h)$$

so

$$P_{in}(t+h)-P_{in}(t)=-\lambda_i h P_{in}(t)+\lambda_i h P_{i+1,n}(t)+o(h).$$

Then

$$\frac{P_{in}(t+h)-P_{in}(t)}{h}=-\lambda_i P_{in}(t)+\lambda_i P_{i+1,n}(t)+\frac{o(h)}{h}$$

and so

$$\lim_{h\to 0}\frac{P_{in}(t+h)-P_{in}(t)}{h}=P'_{in}(t)=-\lambda_i P_{in}(t)+\lambda_i P_{i+1,n}(t). \tag{11}$$

Equation (11) is a backward equation for a pure birth process that agrees with what we knew from the general theory.

Birth and Death Processes

In the birth and death model, the states are 0, 1, 2, …, and transitions are allowed from i to $i+1$ or $i-1$ except 0 can transition only to 1. We let λ_i be the birth rate when the system is in state i and μ_i be the death rate when the system is in state i. We define $\mu_0 = 0$.

For $i \geq 0$, we postulate that state i transitions to state $i+1$ in h units of time with probability $\lambda_i h + o(h)$, and for $i \geq 1$, state i transitions to state $i-1$ in h units of time with probability $\mu_i h + o(h)$, where h is small.

The probability of more than one transition in h units of time is $o(h)$. So we have

$$P(X_{t+h}=n+1 \mid X_t=n)=\lambda_n h+o(h), \quad n\geq 0$$

$$P(X_{t+h}=n-1 \mid X_t=n)=\mu_n h+o(h), \quad n\geq 1$$

$$P(X_{t+h}=n \mid X_t=n)=1-\lambda_n h-\mu_n h+o(h)$$

$$P(X_{t+h}=k \mid X_t=n)=o(h), \quad \text{if } k\neq n, n+1, n-1.$$

Thus, the rate of transition is $e^{-\lambda_n-\mu_n}$, and by Theorem 4.6,

$$P(n,n+1)=\frac{\lambda_n}{\lambda_n+\mu_n}, \quad P(n,n-1)=\frac{\mu_n}{\lambda_n+\mu_n}.$$

A birth–death process is irreducible if and only if $\lambda_n > 0$ for $n \geq 0$ and $\mu_n > 0$ for $n \geq 1$. These are the only birth–death processes that we consider.

Forward Equations for Birth–Death Processes

The comments about notation for the forward equations for the birth process are valid for the birth–death process also.

We recall the forward equations for the general case,

$$P'_{ij}(t) = -\lambda(j)P_{ij}(t) + \sum_{k:k\neq j}\left[\lambda(k)p(k,j)\right]P_{ik}(t).$$

In a birth–death process, the only possible transitions are $n \to n+1$ and $n \to n-1$. Thus,

$$\begin{aligned} P'_{ij} &= -\lambda(j)P_{ij}(t) + \lambda(j-1)p(j-1,j)P_{i(j-1)}(t) \\ &\quad + \lambda(j+1)p(j+1,j)P_{i(j+1)}(t). \end{aligned}$$

Now

$$p(j-1,j) = \frac{\lambda_{j-1}}{\lambda_{j-1}+\mu_{j-1}},$$

$$p(j+1,j) = \frac{\lambda_{j+1}}{\lambda_{j+1}+\mu_{j+1}},$$

and the change rate in state $(j-1)$ is $(\lambda_{j-1}+\mu_{j-1})$, and the change rate in state $(j+1)$ is $(\lambda_{j+1}+\mu_{j+1})$. Thus, in the general formula,

$$\lambda(j-1) = \lambda_{j-1}+\mu_{j-1} \quad \text{and} \quad \lambda(j+1) = \lambda_{j+1}+\mu_{j+1}$$

so we have

$$\begin{aligned} P'_{ij} &= -\lambda(j)P_{ij}(t) + \lambda(j-1)p(j-1,j)P_{i(j-1)}(t) + \lambda(j+1)p(j+1,j)P_{i(j+1)}(t) \\ &= -\lambda(j)P_{ij}(t) + (\lambda_{j-1}+\mu_{j-1})\left(\frac{\lambda_{j-1}}{\lambda_{j-1}+\mu_{j-1}}\right)P_{i(j-1)}(t) \\ &\quad + (\lambda_{j+1}+\mu_{j+1})\left(\frac{\mu_{j+1}}{\lambda_{j+1}+\mu_{j+1}}\right)P_{i(j+1)}(t) \\ &= -\lambda(j)P_{ij}(t) + \lambda_{j-1}P_{i(j-1)}(t) + \mu_{j+1}P_{i(j+1)}(t). \end{aligned}$$

This can also be written as

$$P'_{ij} = -(\lambda_j+\mu_j)P_{ij}(t) + \lambda_{j-1}P_{i(j-1)}(t) + \mu_{j+1}P_{i(j+1)}(t).$$

As with the birth process, one usually assumes the initial condition $P_i(0) = 1$ and $P_k(0) = 0$ if $k \neq i$ and sets $P_n(t) = P_{in}(t)$ to express this as

$$P_n'(t) = -(\lambda_n + \mu_n)P_n(t) + P_{n-1}(t)\lambda_{n-1} + P_{n+1}(t)\mu_{n+1}.$$

We show the derivation of this formula from first principles.

We compute $P_n(t + h)$, which is the probability of the system being in state n at time $t + h$.

The system will be in state n at time $t + h$ if the following happens:

1. At time t, the system was in state n and no changes occurred between t and $t + h$. That there are no transitions from n in the time between t and $t + h$ occurs with probability $1 - \lambda_n h - \mu_n h$, so this event occurs with probability $P_n(t)[1 - \lambda_n h - \mu_n h]$.
2. At time t, the system was in state $n - 1$ and exactly one change (a birth) occurred between t and $t + h$. The birth occurs with probability, so this event occurs with probability $P_{n-1}(t)\left[\lambda_{n-1}h + o(h)\right]$.
3. At time t, the system was in state $n + 1$ and exactly one change (a death) occurred between t and $t + h$. The death occurs with probability $\mu_{n+1}h + o(h)$, so this event occurs with probability $P_{n+1}(t)\left[\mu_{n+1}h + o(h)\right]$.
4. Any other change requires more than one transition, and this occurs with probability $o(h)$.

Thus,

$$P_n(t+h) = P_n(t)[1 - \lambda_n h - \mu_n h] + P_{n-1}(t)\left[\lambda_{n-1}h + o(h)\right] + P_{n+1}(t)\left[\mu_{n+1}h + o(h)\right]$$

so

$$P_n(t+h) - P_n(t) = -(\lambda_n + \mu_n)P_n(t)h + P_{n-1}(t)\lambda_{n-1}h + P_{n+1}(t)\mu_{n+1}h + o(h).$$

Then

$$\frac{P_n(t+h) - P_n(t)}{h} = -(\lambda_n + \mu_n)P_n(t) + P_{n-1}(t)\lambda_{n-1} + P_{n+1}(t)\mu_{n+1} + \frac{o(h)}{h}$$

and

$$P_n'(t) = \lim_{h \to 0} \frac{P_n(t+h) - P_n(t)}{h} = -(\lambda_n + \mu_n)P_n(t) + P_{n-1}(t)\lambda_{n-1} + P_{n+1}(t)\mu_{n+1}.$$

A guide that is helpful in understanding and modeling processes like birth–death processes is a transition rate diagram. The transition rate diagram for the process we are discussing is shown in Figure 5.2.

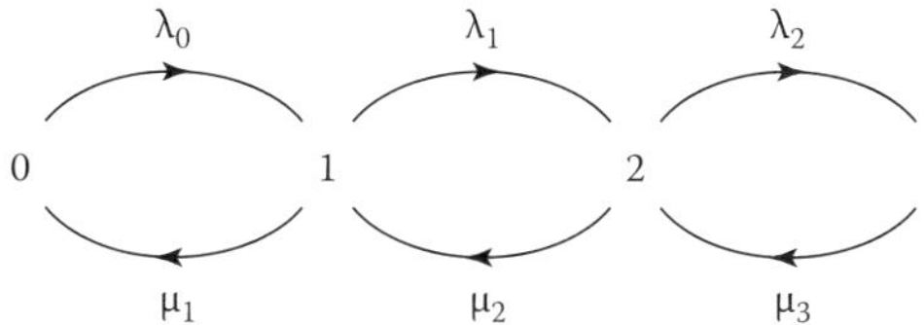

FIGURE 5.2
Transition rate diagram.

We also get that the embedded Markov chain, E, and the infinitesimal generator, G, are given by

$$E = \begin{pmatrix} 0 & \lambda_0 & 0 & 0 & \cdots \\ \mu_1 & 0 & \lambda_1 & 0 & \cdots \\ 0 & \mu_2 & 0 & \lambda_2 & 0 \\ \vdots & \vdots & \vdots & \vdots & \vdots \end{pmatrix}$$

$$G = \begin{pmatrix} -\lambda_0 & \lambda_0 & 0 & 0 & \cdots \\ \mu_1 & -(\lambda_1 + \mu_1) & \lambda_1 & 0 & \cdots \\ 0 & \mu_2 & -(\lambda_1 + \mu_1) & \lambda_2 & 0 \\ \vdots & \vdots & \vdots & \vdots & \vdots \end{pmatrix}.$$

Birth–Death Processes: Backward Equations

The Kolmogorov backward equation for a general process is

$$P'_{ij}(t =) - \lambda(i) P_{ij}(t) + \sum_{k:k \neq i} \left[p(i,k)\lambda(j)h \right] P_{kj}(t).$$

We assume that the only transitions from state n are to $n + 1$ or $n - 1$ and that the exponential rate of transition out of state n is $\lambda_n + \mu_n$ with

$$P(n, n+1) = \frac{\lambda_n}{\lambda_n + \mu_n}, \quad P(n, n-1) = \frac{\mu_n}{\lambda_n + \mu_n}.$$

Thus, the backward equation for the birth–death process is

$$P'_{ij} = (t) = -(\lambda_i + \mu_i) P_{ij}(t) + \left(\frac{\lambda_i}{\lambda_i + \mu_i} \right) (\lambda_i + \mu_i) P_{(i+1)j}(t) + \left(\frac{\mu_i}{\lambda_i + \mu_i} \right) (\lambda_i + \mu_i) P_{(i-1)j}(t)$$

$$= -(\lambda_i + \mu_i) P_{ij}(t) + \lambda_i P_{(i+1)j}(t) + \mu_i P_{(i-1)j}(t).$$

We leave the derivation of the backward equations from first principles as an exercise.

Recurrence and Transience in Birth–Death Processes

Recall that an irreducible Markov chain is recurrent if and only if each state is visited infinitely often with probability 1. Otherwise, the chain is transient. A continuous-time Markov chain is recurrent if and only if the embedded Markov chain is recurrent. Also a recurrent chain in an infinite-state system can be null recurrent or positive recurrent. A recurrent chain is positive recurrent if and only if there is a unique equilibrium distribution. In this section, we give conditions on a birth–death process that determines to which of these classes a chain belongs.

The next results give a way to determine whether a birth–death process is positive recurrent, null recurrent, or transient.

Theorem 5.5

An irreducible birth–death Markov chain with birth rates λ_i and death rates μ_i is positive recurrent if and only if

$$\sum_{n=0}^{\infty} \frac{\lambda_0 \lambda_1 \cdots \lambda_{n-1}}{\mu_1 \mu_2 \cdots \mu_n} = L < \infty.$$

(The $n = 0$ term is defined to be 1.)

In this case, the equilibrium probability distribution is given by $\hat{\pi} = \pi_n$ where

$$\pi_n = \frac{\dfrac{\lambda_0 \lambda_1 \cdots \lambda_{n-1}}{\mu_1 \mu_2 \cdots \mu_n}}{L}.$$

Proof

Recall that the steady state $\hat{\pi} = (\pi_0, \pi_1, \pi_2, \pi_3, \ldots)$ is found by $\hat{\pi}G = \hat{0} = (0,0,0,\ldots)$. Now

$$\hat{\pi}G = (\pi_0, \pi_1, \pi_2, \pi_3, \ldots)\begin{pmatrix} -\lambda_0 & \lambda_0 & 0 & 0 & \cdots \\ \mu_1 & -(\lambda_1+\mu_1) & \lambda_1 & 0 & \cdots \\ 0 & \mu_2 & -(\lambda_2+\mu_2) & \lambda_2 & 0 \\ \vdots & \vdots & \vdots & \vdots & \vdots \end{pmatrix} = \hat{0}$$

yields the following equations:

$$-\lambda_0\pi_0 + \mu_1\pi_1 = 0 \tag{12}$$

$$\lambda_0\pi_0 - (\lambda_1 + \mu_1)\pi_1 + \mu_2\pi_2 = 0 \tag{13}$$

$$\lambda_1\pi_1 - (\lambda_2 + \mu_2)\pi_2 + \mu_3\pi_3 = 0 \tag{14}$$

$$\lambda_2\pi_2 - (\lambda_3 + \mu_3)\pi_3 + \mu_4\pi_4 = 0 \tag{15}$$

$$\vdots$$

These equations enable us to express π_k, $k > 0$ in terms of π_0, λ_j, and μ_j.

From equation (12), we get

$$\mu_1\pi_1 = \lambda_0\pi_0 \quad \text{so} \quad \pi_1 = \pi_0\frac{\lambda_0}{\mu_1}.$$

From equation (13), we get

$$\lambda_0\pi_0 + \mu_2\pi_2 = (\lambda_1 + \mu_1)\pi_1 = (\lambda_1 + \mu_1)\pi_0\frac{\lambda_0}{\mu_1} = \frac{\lambda_0\lambda_1}{\mu_1}\pi_0 + \lambda_0\pi_0$$

so

$$\mu_2\pi_2 = \frac{\lambda_0\lambda_1}{\mu_1}\pi_0 \quad \text{and} \quad \pi_2 = \frac{\lambda_0\lambda_1}{\mu_1\mu_2}\pi_0.$$

From equation (14), we get

$$\lambda_1\pi_1 + \mu_3\pi_3 = (\lambda_2 + \mu_2)\pi_2 \quad \text{so} \quad \mu_3\pi_3 = -\lambda_1\pi_1 + (\lambda_2 + \mu_2)\pi_2$$

$$= -\lambda_1\left(\pi_0\frac{\lambda_0}{\mu_1}\right) + (\lambda_2 + \mu_2)\left(\frac{\lambda_0\lambda_1}{\mu_1\mu_2}\pi_0\right) = \frac{\lambda_0\lambda_1\lambda_2}{\mu_1\mu_2}\pi_0$$

so

$$\pi_3 = \frac{\lambda_0\lambda_1\lambda_2}{\mu_1\mu_2\mu_3}\pi_0.$$

The pattern

$$\pi_n = \frac{\lambda_0\,\lambda_1\ldots\lambda_{n-1}}{\mu_1\,\mu_2\ldots\mu_n}\pi_0$$

can be verified by induction.

Suppose that

$$\sum_{n=0}^{\infty} \frac{\lambda_0 \lambda_1 \ldots \lambda_{n-1}}{\mu_1 \mu_2 \ldots \mu_n} = L < \infty.$$

If $\lambda_i > 0$ for $i \geq 0$, and $\mu_i > 0$, for $i \geq 1$, then all states communicate and the chain is irreducible. If

$$\pi_0 = \frac{1}{L}$$

and for $n \geq 1$

$$\pi_n = \frac{\dfrac{\lambda_0 \lambda_1 \ldots \lambda_{n-1}}{\mu_1 \mu_2 \ldots \mu_n}}{L},$$

then $\pi_n > 0$ and

$$\sum_{n=0}^{\infty} \pi_n = \frac{1}{L} \sum_{n=0}^{\infty} \frac{\lambda_0 \lambda_1 \cdots \lambda_{n-1}}{\mu_1 \mu_2 \cdots \mu_n} = 1.$$

Thus, $\hat{\pi}$ is the unique equilibrium probability distribution with positive entries.

To determine when such a process is transient, we use the following result, which is Theorem 3.9.

Theorem 5.6

Let X_n be an irreducible Markov chain with state space S. Fix $s_0 \in S$. Let

$$\alpha(j) = P(X_n = s_0 \quad \text{for some } n \geq 0 \mid X_n = j).$$

Then X_n is transient if and only if there is a unique solution to $\alpha : S \to \mathbb{R}$ satisfying

(1) $0 \leq \alpha(j) \leq 1$
(2) $\alpha(s_0) = 1, \quad \inf\{\alpha(j) \mid j \in S\} = 0$
(3) For $x \neq s_0$,

$$\alpha(x) = \sum_{y \in S} p(x,y)\alpha(y)$$

The intuition of the theorem is this. Take $s_0 = 0$ and let $\alpha(j)$ be the probability that starting in state j the process ever reaches state 0. If the chain is transient, then $\alpha(j) \to 0$ as $j \to \infty$.

In the proof of the following theorem, we assume that $\mu_i + \lambda_i = 1$, as they must if we are to embed the discrete-time process into a continuous process. If one is only interested in the discrete-time model, it is possible to take $\mu_i > 0$ and $\lambda_i > 0$ but $\mu_i + \lambda_i < 1$ and set $P(i, i) = 1 - \mu_i - \lambda_i$. The theorem is still true in this case, and the proof is virtually identical.

Theorem 5.7

An irreducible birth–death Markov chain is transient if and only if

$$\sum_{n=1}^{\infty} \frac{\mu_1 \ldots \mu_n}{\lambda_1 \ldots \lambda_n} < \infty.$$

(Note the difference between

$$\frac{\mu_1 \ldots \mu_n}{\lambda_1 \ldots \lambda_n}$$

of this theorem and

$$\frac{\lambda_0 \, \lambda_1 \ldots \lambda_{n-1}}{\mu_1 \, \mu_2 \ldots \mu_n}$$

of Theorem 5.5.)

Proof

Let 0 be s_0 in the previous theorem. So $\alpha(0) = 1$. The probability of a death in state n is

$$\frac{\mu_n}{\mu_n + \lambda_n}, \quad n \geq 1$$

and the probability of a birth in state n is

$$\frac{\lambda_n}{\mu_n + \lambda_n} \quad n \geq 0.$$

Now $\alpha(n)$ is the probability that beginning in state n the process ever reaches state 0. So

$$\begin{aligned}\alpha(n) &= P(X_k = 0 \quad \text{for some } k \geq 1 \mid X_0 = n) \\ &= \sum_j P(X_k = 0 \quad \text{for some } k \geq 1, X_1 = j \mid X_0 = n) \\ &= \sum_j P(X_k = 0 \quad \text{for some } k \geq 1, \mid X_1 = j) P(X_1 = j \mid X_0 = n).\end{aligned}$$

Now if $n \geq 1$, then

$$p(n,n-1) = \frac{\mu_n}{\mu_n + \lambda_n}, \quad p(n,n+1) = \frac{\lambda_n}{\mu_n + \lambda_n} \quad \text{and} \quad p(n,x) = 0 \text{ otherwise}$$

so we have

$$\alpha(n) = \frac{\mu_n}{\mu_n + \lambda_n}\alpha(n-1) + \frac{\lambda_n}{\mu_n + \lambda_n}\alpha(n+1) = \mu_n\alpha(n-1) + \lambda_n\alpha(n+1)$$

if $\mu_n + \lambda_n = 1$.

This fulfills the condition

$$\alpha(x) = \sum_{y \in S} p(x,y)\alpha(y).$$

We proceed to find a solution for $\alpha(n + 1)$. We have

$$\alpha(n)(\mu_n + \lambda_n) - \alpha(n+1)\lambda_n = \alpha(n-1)\mu_n$$

so

$$\left[\alpha(n) - \alpha(n+1)\right]\lambda_n = \left[\alpha(n-1) - \alpha(n)\right]\mu_n$$

and

$$\alpha(n) - \alpha(n+1) = \frac{\mu_n}{\lambda_n}\left[\alpha(n-1) - \alpha(n)\right].$$

(Here we have allowed for the more general case $\mu_i + \lambda_i < 1$.)

Likewise,

$$\alpha(n-1)-\alpha(n)=\frac{\mu_{n-1}}{\lambda_{n-1}}\left[\alpha(n-2)-\alpha(n-1)\right]$$

so

$$\alpha(n)-\alpha(n+1)=\frac{\mu_n}{\lambda_n}\left[\alpha(n-1)-\alpha(n)\right]=\frac{\mu_n\ \mu_{n-1}}{\lambda_n\ \lambda_{n-1}}\left[\alpha(n-2)-\alpha(n-1)\right].$$

Continuing the process yields

$$\alpha(n)-\alpha(n+1)=\frac{\mu_n\ \mu_{n-1}\ldots\mu_1}{\lambda_n\ \lambda_{n-1}\ldots\lambda_1}\left[\alpha(0)-\alpha(1)\right].$$

Now

$$\alpha(n+1)=\left[\alpha(n+1)-\alpha(0)\right]+\alpha(0)$$

and

$$\begin{aligned}\left[\alpha(n+1)-\alpha(0)\right]&=\left[\alpha(n+1)-\alpha(n)\right]+\left[\alpha(n)-\alpha(n-1)\right]+\cdots+\left[\alpha(1)-\alpha(0)\right]\\&=\frac{\mu_n\ldots\mu_1}{\lambda_n\ldots\lambda_1}\left[\alpha(1)-\alpha(0)\right]+\frac{\mu_{n-1}\ldots\mu_1}{\lambda_{n-1}\ldots\lambda_1}\left[\alpha(1)-\alpha(0)\right]+\cdots\\&\quad+\frac{\mu_1}{\lambda_1}\left[\alpha(1)-\alpha(0)\right]+\left[\alpha(1)-\alpha(0)\right]\\&=\left[\alpha(1)-\alpha(0)\right]\sum_{j=0}^{n}\frac{\mu_j\ldots\mu_1}{\lambda_j\ldots\lambda_1}\end{aligned}$$

where we take the $j = 0$ term to be 1.

Now $\alpha(0) = 1$, so

$$\begin{aligned}\alpha(n+1)&=\left[\alpha(n+1)-\alpha(0)\right]+\alpha(0)=\left[\alpha(1)-\alpha(0)\right]\sum_{j=0}^{n}\frac{\mu_j\ldots\mu_1}{\lambda_j\ldots\lambda_1}+\alpha(0)\\&=\left[\alpha(1)-1\right]\sum_{j=0}^{n}\frac{\mu_j\ldots\mu_1}{\lambda_j\ldots\lambda_1}+1.\end{aligned}$$

If the series

$$\sum_{j=0}^{n} \frac{\mu_j \ldots \mu_1}{\lambda_j \ldots \lambda_1}$$

diverges, then it is impossible for $\alpha(n) \to 0$ as $n \to \infty$. In this case, by Theorem 5.5, the chain cannot be transient.

Suppose the aforementioned series converges. Let

$$L = \sum_{j=0}^{\infty} \frac{\mu_j \ldots \mu_1}{\lambda_j \ldots \lambda_1}$$

and let

$$\alpha(1) = \frac{L-1}{L} = \frac{\sum_{j=1}^{\infty} \frac{\mu_j \ldots \mu_1}{\lambda_j \ldots \lambda_1}}{\sum_{j=0}^{\infty} \frac{\mu_j \ldots \mu_1}{\lambda_j \ldots \lambda_1}}.$$

Then

$$\alpha(n+1) = [\alpha(1) - 1] \sum_{j=0}^{n} \frac{\mu_j \ldots \mu_1}{\lambda_j \ldots \lambda_1} + 1 = \left[\frac{\sum_{j=1}^{\infty} \frac{\mu_j \ldots \mu_1}{\lambda_j \ldots \lambda_1}}{\sum_{j=0}^{\infty} \frac{\mu_j \ldots \mu_1}{\lambda_j \ldots \lambda_1}} - 1 \right] \left[\sum_{j=0}^{n} \frac{\mu_j \ldots \mu_1}{\lambda_j \ldots \lambda_1} + 1 \right]$$

$$= \left[\frac{\sum_{j=1}^{\infty} \frac{\mu_j \ldots \mu_1}{\lambda_j \ldots \lambda_1} - \sum_{j=0}^{\infty} \frac{\mu_j \ldots \mu_1}{\lambda_j \ldots \lambda_1}}{\sum_{j=0}^{\infty} \frac{\mu_j \ldots \mu_1}{\lambda_j \ldots \lambda_1}} \right] \left[\sum_{j=0}^{n} \frac{\mu_j \ldots \mu_1}{\lambda_j \ldots \lambda_1} \right] + 1.$$

Now the $j = 0$ term of

$$\sum_{j=0}^{\infty} \frac{\mu_j \ldots \mu_1}{\lambda_j \ldots \lambda_1}$$

is 1, so

$$\sum_{j=1}^{\infty} \frac{\mu_j \ldots \mu_1}{\lambda_j \ldots \lambda_1} - \sum_{j=0}^{\infty} \frac{\mu_j \ldots \mu_1}{\lambda_j \ldots \lambda_1} = -1$$

and

$$\left[\frac{\sum_{j=1}^{\infty}\frac{\mu_j\ldots\mu_1}{\lambda_j\ldots\lambda_1}-\sum_{j=0}^{\infty}\frac{\mu_j\ldots\mu_1}{\lambda_j\ldots\lambda_1}}{\sum_{j=0}^{\infty}\frac{\mu_j\ldots\mu_1}{\lambda_j\ldots\lambda_1}}\right]\left[\sum_{j=0}^{n}\frac{\mu_j\ldots\mu_1}{\lambda_j\ldots\lambda_1}\right]+1$$

$$=\left[\frac{-1}{\sum_{j=0}^{\infty}\frac{\mu_j\ldots\mu_1}{\lambda_j\ldots\lambda_1}}\right]\left[\sum_{j=0}^{n}\frac{\mu_j\ldots\mu_1}{\lambda_j\ldots\lambda_1}\right]+\frac{\sum_{j=0}^{\infty}\frac{\mu_j\ldots\mu_1}{\lambda_j\ldots\lambda_1}}{\sum_{j=0}^{\infty}\frac{\mu_j\ldots\mu_1}{\lambda_j\ldots\lambda_1}}$$

$$=\frac{\sum_{j=n+1}^{\infty}\frac{\mu_j\ldots\mu_1}{\lambda_j\ldots\lambda_1}}{\sum_{j=0}^{\infty}\frac{\mu_j\ldots\mu_1}{\lambda_j\ldots\lambda_1}}.$$

Since

$$\sum_{j=0}^{\infty}\frac{\mu_j\ldots\mu_1}{\lambda_j\ldots\lambda_1}$$

converges, then

$$\sum_{j=n+1}^{\infty}\frac{\mu_j\ldots\mu_1}{\lambda_j\ldots\lambda_1}\to 0 \text{ as } n\to\infty$$

so $\alpha(n+1)\to 0$ as $n\to\infty$. Thus, by Theorem 5.6, the Markov chain is transient.

By Theorems 5.5 and 5.7, we know exactly when a birth–death process is positive recurrent and when a birth death process is transitive. If we have a birth–death process that meets neither of these criteria, then it must be null recurrent. We will see an application of this in the section on queuing systems.

Example

A much studied group of examples in mathematical biology is modeling infectious diseases. Probably the most common way of doing this is via differential equations, but the method of Markov chains also has its advantages. The two simplest models are the SIS model (standing for susceptible–infectious–susceptible) and the SIR model (susceptible–infectious–recovered). We will formulate the Markov chain associated with each model.

SIS model

We assume that a population has a fixed number of individuals that we denote N. An infection is introduced into the population through a small number of its members. In this model, an individual is either susceptible or infected (and infected individuals are infectious, that is, capable of passing the disease to susceptible members through contact). When a member recovers, he rejoins the susceptible group. We let

$$S(t) = \text{the number of members that are susceptible at time } t$$

$$I(t) = \text{the number of members that are infected at time } t.$$

The disease is passed with probability β when an infected member comes into contact with a susceptible member. We let γ denote the recovery rate. We also have

$$S(t) + I(t) = N.$$

We let the state space be the number of infected individuals, so the state space is

$$\{0, 1, \ldots, N\}.$$

Let $G = \big(g(i,j)\big)$ be the infinitesimal generator. Then $g(0, k) = 0$ if $k \neq 0$ since the infection cannot be passed if everyone is healthy.

The number of contacts depends on the number of contacts that infectious and susceptible members have. It can be shown that the rate of the number of contacts is proportional to $I(t)S(t)$.

So

$$g(k, k+1) = \beta k \frac{(N-k)}{N}, \quad k = 0, \ldots, N-1,$$

$$g(k, k-1) = \gamma k; \quad k = 1, 2, \ldots, N.$$

For $N = 3$, G is

$$G = \begin{pmatrix} 0 & 0 & 0 & 0 \\ \gamma & -(\gamma + 2\beta) & 2\beta & 0 \\ 0 & 2\gamma & -(2\gamma + \beta) & \beta \\ 0 & 0 & 3\gamma & -3\gamma \end{pmatrix}.$$

SIR model

We again assume that a population has a fixed number of individuals that we denote N and that an infection is introduced into the population through a small number of its members. In this model, a member that recovers goes to a third category called the recovered group. Members of this group have recovered from the disease and cannot be reinfected. An example of a disease where the SIS model would be appropriate is a sexually transmitted disease. An example of a disease where the SIR model is more accurate is chicken pox.

We let

$$S(t) = \text{the number of members that are susceptible at time } t$$

$$I(t) = \text{the number of members that are infected at time } t$$

$$R(t) = \text{the number of members that have recovered at time } t.$$

In this model, the state space is necessarily more complicated because knowing only the number of infected individuals does not fully specify the system. If we know the number of members in each of two of the states, then the third state is known. We let the state space be

$$S = \{(s,i) : 0 \le s, i \le N, 0 \le s + i \le N\}$$

where

s is the number of susceptible

i is the number of infected

Consider the possible transitions of a state (s, i). One possibility is that a susceptible could become infected. In this case, state (s, i) has transitioned to state $(s - 1, i + 1)$. This occurs at the rate

$$g((s,i),(s-1,i+1)) = \beta i \frac{s}{N}.$$

The other way state (s, i) could change is that an infected member could recover. In this case, state (s, i) has transitioned to state $(s, i - 1)$. This occurs at the rate

$$g((s,i),(s,i-1)) = \gamma i.$$

In the SIR model, to apply the techniques for Markov chains that we have developed, we need to associate each state with a unique integer. To develop the intuition for a convenient association, consider the case where $N = 3$.

Note that any state with $i = 0$ is an absorbing state, so for the purposes of specifying the transition matrix it will be convenient to list those states first. One possibility is

$$(3,0) \to 1, (2,0) \to 2, (1,0) \to 3, (0,0) \to 4,$$

$$(2,1) \to 5, (1,1) \to 6, (0,1) \to 7,$$

$$(1,2) \to 8, (0,2) \to 9,$$

$$(0,3) \to 10.$$

We determine the infinitesimal generator for this model.

Note that there is no possibility of moving out of a state $(s, 0)$. In terms of our numbering of the states of the infinitesimal matrix, states

1 through 4 have $g(i,j) = 0$ for $i = 1, 2, 3, 4$. If we were listing the transition matrix, we would have $P(i,i) = 1$ for $i = 1, 2, 3, 4$. The original states that have a nonzero transition rate, the corresponding renumbered states, and their transition rate are presented in the following:

Transition of States (Original Numbering)	**Transition of States (New Numbering)**	**Transition Rate**
$(2,1) \to (2,0)$	$5 \to 2$	$\gamma \cdot 1$
$(2,1) \to (1,2)$	$5 \to 8$	$\frac{\beta}{3} \cdot 2 \cdot 1$
$(1,1) \to (0,2)$	$6 \to 9$	$\frac{\beta}{3} \cdot 1 \cdot 1$
$(1,1) \to (1,0)$	$6 \to 3$	$\gamma \cdot 1$
$(0,1) \to (0,0)$	$7 \to 4$	$\gamma \cdot 1$
$(1,2) \to (0,3)$	$8 \to 10$	$\frac{\beta}{3} \cdot 1 \cdot 2$
$(1,2) \to (1,1)$	$8 \to 6$	$\gamma \cdot 2$
$(0,2) \to (0,1)$	$9 \to 7$	$\gamma \cdot 2$
$(0,3) \to (0,2)$	$10 \to 9$	$\gamma \cdot 3.$

For this model,

$$G = \begin{pmatrix} 0 & 0 & 0 & 0 & 0 & 0 & 0 & 0 & 0 & 0 \\ 0 & 0 & 0 & 0 & 0 & 0 & 0 & 0 & 0 & 0 \\ 0 & 0 & 0 & 0 & 0 & 0 & 0 & 0 & 0 & 0 \\ 0 & 0 & 0 & 0 & 0 & 0 & 0 & 0 & 0 & 0 \\ 0 & \gamma & 0 & 0 & -\left(\gamma + \frac{2\beta}{3}\right) & 0 & 0 & \frac{2\beta}{3} & 0 & 0 \\ 0 & 0 & \gamma & 0 & 0 & 0-\left(\gamma + \frac{\beta}{3}\right) & 0 & 0 & \frac{\beta}{3} & 0 \\ 0 & 0 & 0 & \gamma & 0 & 0 & -\left(\gamma + \frac{2\beta}{3}\right) & 0 & 0 & 0 \\ 0 & 0 & 0 & 0 & 0 & 2\gamma & 0 & -\left(2\gamma + \frac{2\beta}{3}\right) & 0 & \frac{2\beta}{3} \\ 0 & 0 & 0 & 0 & 0 & 0 & 2\gamma & 0 & -2\gamma & 0 \\ 0 & 0 & 0 & 0 & 0 & 0 & 0 & 0 & 3\gamma & -3\gamma \end{pmatrix}.$$

Queuing Models

In a queuing model, there is a group of *customers* in a line waiting to be served. We hypothesize the arrival rate of the customers and the time required to serve a customer. The models are often denoted $A/S/k$ where the choice of A gives the distribution of the arrival rate of the customers, S the distribution of service time, and k the number of servers. A common arrival distribution is a Poisson distribution, and it is designated by M and a common service rate has an exponential distribution that is also designated by M. The letter M is representative of Markovian for the memoryless character of the distribution. The other possibility is if the arrival or service times is some other distribution, and this is designated by G, for general.

In analyzing a queuing model as a Markov process, X_t is the number of customers in the system at time t. The number of customers in the system is the sum of the number waiting in line plus the number being served.

M/M/1 Model

As a first example, we consider the M/M/1 queuing model. This model has important features in common with the birth-death model. An arrival of a customer is analogous to a birth (a new member is entering the population) and completing service is analogous to a death (a member is leaving the population). We assume that the arrivals follow a Poisson distribution with rate λ and the service time is exponential with μ. We then have exactly the same equations governing the dynamics as a birth–death process where $\lambda_i = \lambda$ and $\mu_i = \mu$ for every i.

Earlier, we showed that the birth–death process has a unique equilibrium state exactly when

$$\sum_{n=0}^{\infty} \frac{\lambda_0 \lambda_1 \ldots \lambda_{n-1}}{\mu_1 \mu_2 \ldots \mu_n}$$

converges. Since in our case $\lambda_i = \lambda$ and $\mu_i = \mu$ for every I, this reduces to

$$\sum_{n=0}^{\infty} \frac{\lambda^n}{\mu^n} = \sum_{n=0}^{\infty} \left(\frac{\lambda}{\mu}\right)^n$$

and this series converges exactly when

$$\frac{\lambda}{\mu} < 1; \quad \text{that is, } \lambda < \mu.$$

We also found that the equilibrium state is given by

$$\pi_n = \frac{\dfrac{\lambda_0 \lambda_1 \ldots \lambda_{n-1}}{\mu_1 \mu_2 \ldots \mu_n}}{\sum_{n=0}^{\infty} \dfrac{\lambda_0 \lambda_1 \ldots \lambda_{n-1}}{\mu_1 \mu_2 \ldots \mu_n}} = \frac{\left(\dfrac{\lambda}{\mu}\right)^n}{\sum_{n=0}^{\infty} \left(\dfrac{\lambda}{\mu}\right)^n} = \frac{\left(\dfrac{\lambda}{\mu}\right)^n}{\dfrac{1}{1-\left(\dfrac{\lambda}{\mu}\right)}} = \left(1-\frac{\lambda}{\mu}\right)\left(\frac{\lambda}{\mu}\right)^n.$$

In analyzing a queue, here are some of the parameters of likely interest and their formulas:

The expected number of customers in the system is

$$L=\sum_{n=0}^{\infty}n\left(1-\frac{\lambda}{\mu}\right)\left(\frac{\lambda}{\mu}\right)^{n}=\left(1-\frac{\lambda}{\mu}\right)\sum_{n=0}^{\infty}n\left(\frac{\lambda}{\mu}\right)^{n}=\sum_{n=0}^{\infty}\left(1-\frac{\lambda}{\mu}\right)\frac{\left(\frac{\lambda}{\mu}\right)}{\left(1-\left(\frac{\lambda}{\mu}\right)\right)^{2}}=\frac{\lambda}{\mu-\lambda}.$$

$$\lim_{\lambda\uparrow\mu}L=\infty.$$

The probability that an entering customer must enter the queue (as opposed to being served immediately) is

$$\lim_{t\to\infty}P\big(X(t)\geq 1\big)=\sum_{n=1}^{\infty}\pi(0)\left(\frac{\lambda}{\mu}\right)^{n}=\frac{\lambda}{\mu}.$$

The expected number of customers in the queue is

$$L_Q=L-\frac{\lambda}{\mu}=\frac{\lambda}{\mu-\lambda}-\frac{\lambda}{\mu}=\frac{\lambda^2}{\mu(\mu-\lambda)}.$$

The expected time in the system is

$$T=\frac{L}{\lambda}=\frac{\frac{\lambda}{\mu-\lambda}}{\lambda}=\frac{1}{\mu-\lambda}.$$

The waiting time in the queue, W_Q, is the total waiting time minus the service time, so

$$W_Q=T-\frac{1}{\mu}=\frac{1}{\mu-\lambda}-\frac{1}{\mu}=\frac{\mu-(\mu-\lambda)}{(\mu-\lambda)\mu}=\frac{\lambda}{\mu}\cdot\frac{1}{(\mu-\lambda)}=\frac{\lambda}{\mu(\mu-\lambda)}.$$

The transition rate diagram for the M/M/1 model is shown in Figure 5.3.

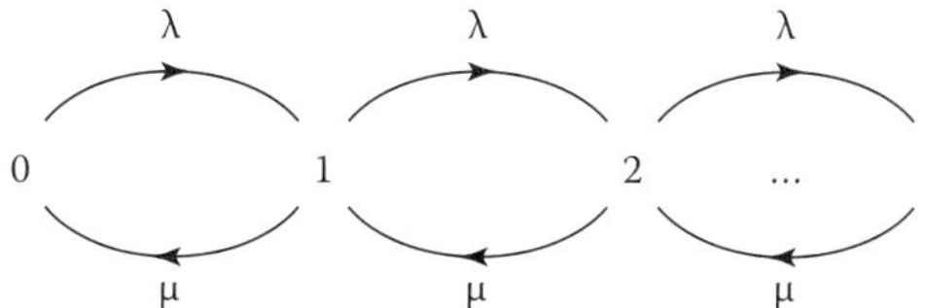

FIGURE 5.3
Transition rate diagram for M/M/1 model.

The infinitesimal generator is

$$G = \begin{pmatrix} -\lambda & \lambda & 0 & 0 & 0 & \cdots \\ \mu & -(\lambda+\mu) & \lambda & 0 & 0 & \cdots \\ 0 & \mu & -(\lambda+\mu) & \lambda & 0 & \cdots \\ \vdots & \vdots & \vdots & \vdots & \vdots & \vdots \end{pmatrix}$$

and the transition matrix for the embedded Markov chain is

$$P = \begin{pmatrix} 0 & \lambda & 0 & 0 & 0 & \cdots \\ \mu & 0 & \lambda & 0 & 0 & \cdots \\ 0 & \mu & 0 & \lambda & 0 & \cdots \\ \vdots & \vdots & \vdots & \vdots & \vdots & \vdots \end{pmatrix}.$$

M/M/1/K Queue: One Server (Size of the Queue Is Limited)

In this scenario, we have one server and the queue has finite capacity. In many problems, this is more realistic than having an infinite queue. An example is a barber shop that has one barber and two chairs for waiting customers. In this case, if a potential customer enters the shop and sees that there are two customers waiting, then that customer will leave the shop and will not become part of the queue. Thus, we have an additional question: How likely are we to lose a customer? We do a numerical example before presenting the abstract case.

Suppose we have the aforementioned barber shop and customers enter the shop at the rate $\lambda = 2$ and they are serviced at the rate $\mu = 3$. The states of the system are the number of customers in the shop, which are 0, 1, 2, and 3. The transition rate diagram of the process is shown in Figure 5.4.

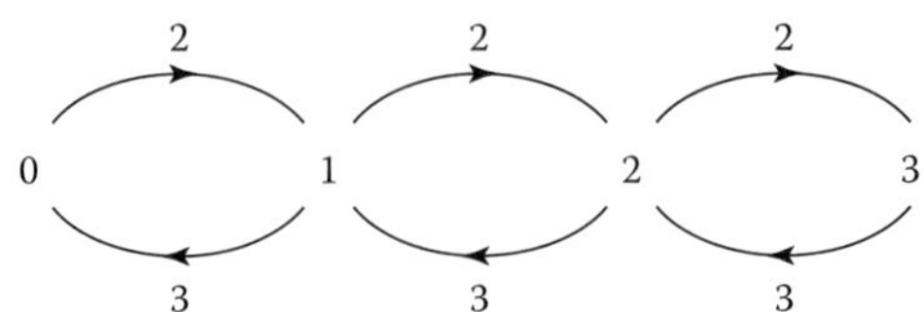

FIGURE 5.4
Transition rate diagram for M/M/3 model.

And the infinitesimal generator is

$$G = \begin{pmatrix} -2 & 2 & 0 & 0 \\ 3 & -5 & 2 & 0 \\ 0 & 3 & -5 & 2 \\ 0 & 0 & 3 & -3 \end{pmatrix}.$$

The equilibrium state is the probability vector $\pi = (\pi_0, \pi_1, \pi_2, \pi_3)$ for which

$$\pi G = (0,0,0,0).$$

Solving $\pi G = (0,0,0,0)$ yields

$$\pi_1 = \frac{2}{3}\pi_0, \quad \pi_2 = \frac{4}{9}\pi_0 = \left(\frac{2}{3}\right)^2 \pi_0, \quad \pi_3 = \frac{8}{27}\pi_0 = \left(\frac{2}{3}\right)^3 \pi_0.$$

Since π is a probability vector,

$$1 = \pi_0 + \pi_1 + \pi_2 + \pi_3 = \pi_0 + \frac{2}{3}\pi_0 + \left(\frac{2}{3}\right)^2 \pi_0 + \left(\frac{2}{3}\right)^3 \pi_0$$

so

$$\pi_0 = \frac{1}{1 + \frac{2}{3} + \left(\frac{2}{3}\right)^2 + \left(\frac{2}{3}\right)^3}.$$

Note that

$$1 + \frac{2}{3} + \left(\frac{2}{3}\right)^2 + \left(\frac{2}{3}\right)^3 = \frac{1 - \frac{2}{3}}{1 - \left(\frac{2}{3}\right)^4}.$$

Thus, the equilibrium state is

$$\pi=\left(\frac{1}{1+\frac{2}{3}+\left(\frac{2}{3}\right)^{2}+\left(\frac{2}{3}\right)^{3}},\frac{\left(\frac{2}{3}\right)^{1}}{1+\frac{2}{3}+\left(\frac{2}{3}\right)^{2}+\left(\frac{2}{3}\right)^{3}},\frac{\left(\frac{2}{3}\right)^{2}}{1+\frac{2}{3}+\left(\frac{2}{3}\right)^{2}+\left(\frac{2}{3}\right)^{3}},\frac{\left(\frac{2}{3}\right)^{3}}{1+\frac{2}{3}+\left(\frac{2}{3}\right)^{2}+\left(\frac{2}{3}\right)^{3}}\right).$$

Now consider the general case where the states are 0, 1, …, K, the transition rate from i to $i+1$ is λ, and the transition rate from i to $i-1$ is μ. Since there are finitely many states, we do not require that $\lambda < \mu$ to have a unique equilibrium state. The transition rate diagram is shown in Figure 5.5.

And the infinitesimal generator is

$$G=\begin{pmatrix} -\lambda & \lambda & 0 & 0 & \cdots & \cdots & 0 \\ 0 & \mu & -(\lambda+\mu) & \lambda & \cdots & \cdots & 0 \\ 0 & 0 & \mu & -(\lambda+\mu) & \mu & \cdots & 0 \\ \vdots & \vdots & \vdots & \vdots & \vdots & \vdots & \vdots \\ 0 & \cdots & \cdots & \cdots & 0 & \mu & -\mu \end{pmatrix}.$$

We now have $K+1$ states, 0, 1, …, K.

If we do the computations in this manner, we get the equilibrium state

$$\pi=(\pi_0,\pi_1,\ldots,\pi_K),\quad \pi_j=\left(\frac{\lambda}{\mu}\right)^{j}\pi_0,$$

$$\pi_0=\frac{1}{1+\left(\frac{\lambda}{\mu}\right)+\cdots\left(\frac{\lambda}{\mu}\right)^{K}}=\frac{1-\left(\frac{\lambda}{\mu}\right)}{1-\left(\frac{\lambda}{\mu}\right)^{K+1}}.$$

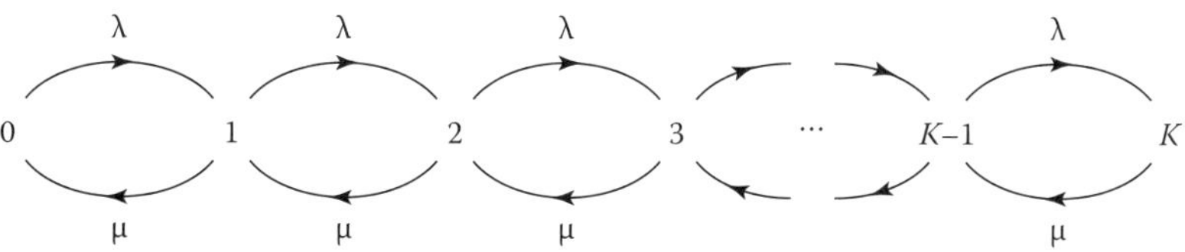

FIGURE 5.5
Transition rate diagram for M/M/1/K model.

The expected number of customers in the queue at time t, $E\left[X(t)\right]$, is

$$E\left[X(t)\right]=\sum_{n=0}^{K} n\pi_n=\sum_{n=0}^{K} n\left(\frac{\lambda}{\mu}\right)^n \pi_0$$

$$=\pi_0\left[\left(\frac{\lambda}{\mu}\right)+2\left(\frac{\lambda}{\mu}\right)^2+\cdots+K\left(\frac{\lambda}{\mu}\right)^K\right]$$

$$=\left[\frac{1-\left(\frac{\lambda}{\mu}\right)}{1-\left(\frac{\lambda}{\mu}\right)^{K+1}}\right]\left[\left(\frac{\lambda}{\mu}\right)+2\left(\frac{\lambda}{\mu}\right)^2+\cdots+K\left(\frac{\lambda}{\mu}\right)^K\right].$$

We leave it as an exercise to show that the last expression is

$$\frac{\left(\frac{\lambda}{\mu}\right)}{1-\left(\frac{\lambda}{\mu}\right)}-\frac{(K+1)\left(\frac{\lambda}{\mu}\right)^{K+1}}{1-\left(\frac{\lambda}{\mu}\right)^{K+1}}.$$

To compute the rate at which customers are turned away, we note that this is the rate at which customers arrive when there are K customers already in the system. Since the arrival rate is λ and the probability there are K customers in the system is π_K, the rate at which customers are turned away is $\lambda\pi_K$.

The rate at which customers actually enter the system, λ_a, is the rate of those that attempt to enter the system minus the rate of those that are turned away. Thus,

$$\lambda_a=\lambda-\lambda\pi_K.$$

Detailed Balance Equations

Detailed balance equations are equations that can be helpful in finding the equilibrium distribution. These equations equate the flow between states in a pairwise fashion. Suppose that i and j are states, P is the transition matrix, and π is the equilibrium distribution. Then the detailed balance equations are

$$\pi(i)P_{ij}=\pi(j)P_{ji}.$$

For a birth–death process, $P_{ij} = 0$ unless $|i - j| = 1$. In this example, we have

$$P_{i,i+1} = \lambda, \quad \text{for } i \geq 0 \quad \text{and} \quad P_{i,i-1} = \mu\,, \quad \text{so } P_{i+i,i} = \mu, \quad \text{for } i \geq 1.$$

Thus,

$$\pi(i)P_{i,i+1} = \lambda\pi(i) \quad \text{and} \quad \pi(i+1)P_{i+i,i} = \mu\pi(i+1),$$

and since

$$\pi(i)P_{i,i+1} = \pi(i+1)P_{i+i,i},$$

we have

$$\lambda\pi(i) = \mu\pi(i+1)$$

or

$$\pi(i+1) = \frac{\lambda}{\mu}\pi(i).$$

So

$$\pi(1) = \frac{\lambda}{\mu}\pi(0),$$

$$\pi(2) = \frac{\lambda}{\mu}\pi(1) = \frac{\lambda}{\mu}\cdot\frac{\lambda}{\mu}\pi(0) = \left(\frac{\lambda}{\mu}\right)^2 \pi(0)$$

$$\pi(3) = \frac{\lambda}{\mu}\pi(2) = \frac{\lambda}{\mu}\left[\left(\frac{\lambda}{\mu}\right)^2 \pi(0)\right] = \left(\frac{\lambda}{\mu}\right)^3 \pi(0)$$

$$\vdots$$

$$\pi(n) = \left(\frac{\lambda}{\mu}\right)^n \pi(0).$$

Thus, we have another way to compute the equilibrium state. Choose $\pi(0)$ so that

$$\sum_{n=0}^{\infty}\left(\frac{\lambda}{\mu}\right)^n \pi(0) = \pi(0)\frac{1}{1-\left(\frac{\lambda}{\mu}\right)} = 1$$

or

$$\pi(0) = 1 - \left(\frac{\lambda}{\mu}\right) = \frac{\mu - \lambda}{\mu}.$$

Then

$$\pi(n) = \left(\frac{\lambda}{\mu}\right)^n \pi(0) = \left(\frac{\lambda}{\mu}\right)^n \left(\frac{\mu - \lambda}{\mu}\right),$$

which is the same as the earlier computation.

Table 5.1 gives a summary of the formulas for the M/M/1 and M/M/1/K queuing models. The quantities are as follows:

P_0 = probability the system is in state 0 at equilibrium

P_n = probability the system is in state n at equilibrium

L = average number of customers in the system

L_q = average number of customers in the queue

W = average waiting time in the system

W_q = average waiting time in the queue

TABLE 5.1

Formulas for the M/M/1 and M/M/1/K Models

	M/M/1	M/M/1/K
P_0	$1-\frac{\lambda}{\mu}$	$\frac{1-\frac{\lambda}{\mu}}{1-\left(\frac{\lambda}{\mu}\right)^{K+1}}$
P_n	$\left(\frac{\lambda}{\mu}\right)^n P_0$	$\left(\frac{\lambda}{\mu}\right)^n P_0$
L	$\frac{\lambda}{\mu-\lambda}$	$\frac{\lambda}{\mu-\lambda}-\frac{(K+1)\left(\frac{\lambda}{\mu}\right)^{K+1}}{1-\left(\frac{\lambda}{\mu}\right)^{K+1}}$
L_q	$\frac{\lambda^2}{\mu(\mu-\lambda)}$	$L-(1-P_0)$
W	$\frac{1}{\mu-\lambda}$	$\frac{L}{\lambda(1-P_K)}$
W_q	$\frac{\lambda}{\mu(\mu-\lambda)} W_q$	$\frac{L_q}{\lambda(1-P_K)}$

M/M/K Model

Now we suppose there are k servers. For the arrival rate, we still have $\lambda_i = \lambda$ for all i, but now the serving rate is

$$\mu_n = \begin{cases} n\mu & \text{if } n < k \\ k\mu & \text{if } n \geq k \end{cases}.$$

The transition rate diagram is shown in Figure 5.6.

And the infinitesimal generator is

$$\begin{pmatrix} -\lambda & \lambda & 0 & 0 & 0 & 0 & & 0 & 0 & 0\ 0 & \cdots & \\ 0 & \mu & -(\mu+\lambda) & \lambda & 0 & 0 & & 0 & 0 & 0\ 0 & \cdots & \\ 0\ 0 & 2\mu & -(2\mu+\lambda) & & \lambda & 0 & & 0 & 0 & 0\ 0 & \cdots & \\ 0\ 0 & 0 & 3\mu & & -(3\mu+\lambda) & \lambda & & 0 & 0 & 0\ 0 & \cdots & \\ \vdots\ \vdots & \vdots & \vdots & \vdots & \vdots & & \vdots\ \vdots & & & \vdots\ \vdots & \vdots & \\ 0\ 0 & 0 & 0 & 0 & 0 & & 0\ k\mu & -(k\mu+\lambda) & & \lambda & \cdots \\ 0\ 0 & 0 & 0 & 0 & 0 & & 0\ 0 & k\mu & & -(k\mu+\lambda) & \lambda \\ \vdots\ \vdots & \vdots & \vdots & \vdots & \vdots & & \vdots\ \vdots & \vdots & & \vdots & \cdots \end{pmatrix}$$

The detailed balance equations for $n < k$ are

$$P_{n,n+1} = \lambda, \ \ P_{n,n-1} = n\mu, \quad \text{so } P_{n+1,n} = (n+1)\mu.$$

Then

$$\pi(n)P_{n,n+1} = \pi(n+1)P_{n+1,n}$$

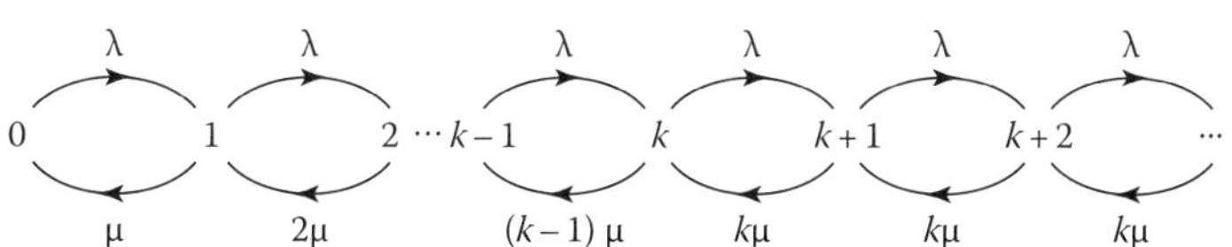

FIGURE 5.6
Transition rate diagram for M/M/K model.

is

$$\lambda\pi(n) = (n+1)\mu\pi(n+1).$$

So

$$\pi(n+1) = \frac{\lambda}{(n+1)\mu}\pi(n).$$

Thus,

$$\pi(1) = \frac{\lambda}{\mu}\pi(0),$$

$$\pi(2) = \frac{\lambda}{2\mu}\pi(1) = \frac{\lambda}{2\mu}\cdot\frac{\lambda}{\mu}\pi(0) = \frac{1}{2}\left(\frac{\lambda}{\mu}\right)^2\pi(0),$$

$$\pi(3) = \frac{\lambda}{3\mu}\pi(2) = \frac{\lambda}{3\mu}\cdot\left[\frac{1}{2}\left(\frac{\lambda}{\mu}\right)^2\pi(0)\right] = \frac{1}{2\cdot 3}\left(\frac{\lambda}{\mu}\right)^3\pi(0),$$

$$\pi(4) = \frac{\lambda}{4\mu}\pi(3) = \frac{\lambda}{4\mu}\cdot\left[\frac{1}{2\cdot 3}\left(\frac{\lambda}{\mu}\right)^3\pi(0)\right] = \frac{1}{2\cdot 3\cdot 4}\left(\frac{\lambda}{\mu}\right)^4\pi(0)$$

$$\vdots$$

$$\pi(n) = \frac{\lambda}{n\mu}\pi(n-1) = \frac{\lambda}{n\mu}\left[\frac{1}{2\cdot 3\cdots(n-1)}\left(\frac{\lambda}{\mu}\right)^{(n-1)}\pi(0)\right] = \frac{1}{n!}\left(\frac{\lambda}{\mu}\right)^n\pi(0).$$

For $n > k$, the detailed balance for $n < k$ equations are

$$P_{n,n+1} = \lambda,\ \ P_{n,n-1} = k\mu, \quad \text{so } P_{n+1,n} = (k+1)\mu.$$

Then

$$\pi(n)P_{n,n+1} = \pi(n+1)P_{n+1,n}$$

is

$$\lambda\pi(n) = (k+1)\mu\pi(n+1).$$

So

$$\pi(n+1) = \frac{\lambda}{(k+1)\mu}\pi(n) = \left(\frac{1}{k}\right)^{n-k}\cdot\frac{1}{k!}\left(\frac{\lambda}{\mu}\right)^n\pi(0).$$

We rewrite

$$\left(\frac{1}{k}\right)^{n-k}\cdot\frac{1}{k!}\left(\frac{\lambda}{\mu}\right)^{n}\pi(0)=k^{k-n}\frac{1}{k!}\left(\frac{\lambda}{\mu}\right)^{n}\pi(0)=\frac{k^k}{k^n}\cdot\frac{1}{k!}\left(\frac{\lambda}{\mu}\right)^{n}\pi(0)=\frac{k^k}{k!}\left(\frac{\lambda}{k\mu}\right)^{n}\pi(0).$$

So we have

$$\pi(n)=\begin{cases}\dfrac{1}{n!}\left(\dfrac{\lambda}{\mu}\right)^{n}\pi(0) & \text{if } n<k\\[2ex] \dfrac{k^k}{k!}\left(\dfrac{\lambda}{k\mu}\right)^{n}\pi(0) & \text{if } n\geq k\end{cases}.$$

To find the explicit value of $\pi(n)$, we must know the value of $\pi(0)$. We have

$$1=\sum_{n=0}^{\infty}\pi(n)=\pi(0)\left[1+\frac{\lambda}{\mu}+\frac{\lambda}{2\mu}\cdot\frac{\lambda}{\mu}+\cdots+\frac{1}{k!}\left(\frac{\lambda}{\mu}\right)^{k}\right]+\pi(0)\left[\sum_{n=k}^{\infty}\frac{k^k}{k!}\left(\frac{\lambda}{k\mu}\right)^{n}\right].$$

We also have

$$\sum_{n=k}^{\infty}\frac{k^k}{k!}\left(\frac{\lambda}{k\mu}\right)^{n}=\frac{k^k}{k!}\sum_{n=k}^{\infty}\left(\frac{\lambda}{k\mu}\right)^{n}.$$

It is common in the literature to let

$$\rho=\frac{\lambda}{k\mu}$$

so that

$$\frac{k^k}{k!}\sum_{n=k}^{\infty}\left(\frac{\lambda}{k\mu}\right)^{n}=\frac{k^k}{k!}\sum_{n=k}^{\infty}\rho^{n}=\frac{k^k}{k!}\cdot\frac{\rho^k}{(1-\rho)}.$$

Thus, we have

$$1=\pi(0)\left[\sum_{n=0}^{k-1}\frac{1}{n!}\left(\frac{\lambda}{\mu}\right)^{n}+\frac{k^k}{k!}\cdot\frac{\rho^k}{(1-\rho)}\right]$$

and so

$$\pi(0)=\left[\sum_{n=0}^{k-1}\frac{1}{n!}\left(\frac{\lambda}{\mu}\right)^n+\frac{k^k}{k!}\cdot\frac{\rho^k}{(1-\rho)}\right]^{-1},$$

which can also be expressed as

$$\pi(0)=\left[\sum_{n=0}^{k-1}\frac{(k\rho)^n}{n!}+\frac{k^k}{k!}\cdot\frac{\rho^k}{(1-\rho)}\right]^{-1}.$$

Next we determine the probability that an entering customer must enter the queue (as opposed to being served immediately). This occurs when there are k or more customers already in the system. Thus, we compute

$$\lim_{t\to\infty}P\big(X(t)\geq k\big)=\sum_{n=k}^{\infty}\pi(0)\frac{k^k}{k!}\rho^n=\pi(0)\frac{k^k}{k!}\sum_{n=k}^{\infty}\rho^n=\pi(0)\frac{k^k}{k!}\frac{\rho^k}{1-\rho},$$

which is the Erlang-C formula.

The expected number of customers in the queue is

$$L_Q=\sum_{n=k}^{\infty}(n-k)\pi(n)=\sum_{n=k}^{\infty}(n-k)\pi(0)\frac{k^k}{k!}\left(\frac{\lambda}{k\mu}\right)^n.$$

Let $j=n-k$. Then $n=j+k$; if $n=k$, then $j=0$, so

$$\begin{aligned}\pi(0)\frac{k^k}{k!}\sum_{n=k}^{\infty}(n-k)\left(\frac{\lambda}{k\mu}\right)^n&=\pi(0)\frac{k^k}{k!}\sum_{j=0}^{\infty}j\left(\frac{\lambda}{k\mu}\right)^{j+k}\\&=\pi(0)\frac{k^k}{k!}\left(\frac{\lambda}{k\mu}\right)^k\sum_{j=0}^{\infty}j\left(\frac{\lambda}{k\mu}\right)^j\\&=\pi(0)\frac{k^k}{k!}\left(\frac{\lambda}{k\mu}\right)^k\sum_{j=1}^{\infty}j\left(\frac{\lambda}{k\mu}\right)^j\\&=\pi(0)\frac{k^k}{k!}\rho^k\sum_{j=1}^{\infty}j\rho^j\\&=\pi(0)\frac{k^k}{k!}\rho^k\frac{\rho}{(1-\rho)^2}.\end{aligned}$$

The expected waiting time in the queue is

$$W_Q = \frac{L_Q}{\lambda}$$

and the expected time within the system is W_Q plus the average time of service, which is

$$W = \frac{L_Q}{\lambda} + \frac{1}{\mu}.$$

The expected number of customers in the system at a given time is

$$L = \lambda W.$$

While these computations were quite tedious, if all we wanted was to find when an equilibrium state exists, we could use

$$\pi_n = \frac{\dfrac{\lambda_{n-1}\dots\lambda_1\lambda_0}{\mu_n\dots\mu_2\mu_1}}{\sum_{n=0}^{\infty}\dfrac{\lambda_{n-1}\dots\lambda_1\lambda_0}{\mu_n\dots\mu_2\mu_1}}$$

with

$$\sum_{n=1}^{\infty}\frac{\lambda_{n-1}\dots\lambda_1\lambda_0}{\mu_n\dots\mu_2\mu_1} = \sum_{n=1}^{k-1}\left(\frac{\lambda}{\mu}\right)^n\cdot\frac{1}{n!} + \sum_{n=k}^{\infty}\left(\frac{\lambda}{\mu}\right)^n\cdot\frac{1}{k^{n-k}\cdot k!}$$

and we see there is a unique equilibrium state when

$$\sum_{n=1}^{\infty}\frac{\lambda_{n-1}\dots\lambda_1\lambda_0}{\mu_n\dots\mu_2\mu_1} = \sum_{n=1}^{k-1}\left(\frac{\lambda}{\mu}\right)^n\cdot\frac{1}{n!} + \sum_{n=k}^{\infty}\left(\frac{\lambda}{\mu}\right)^n\cdot\frac{1}{k^{n-k}\cdot k!}$$

converges. We will show that this happens if and only if $\lambda < k\mu$.

We need only consider

$$\sum_{n=k}^{\infty}\left(\frac{\lambda}{\mu}\right)^n\cdot\frac{1}{k^{n-k}\cdot k!} = \frac{1}{k!}\sum_{n=k}^{\infty}\left(\frac{\lambda}{\mu}\right)^n\cdot\frac{1}{k^{n-k}} = \frac{k^k}{k!}\sum_{n=k}^{\infty}\left(\frac{\lambda}{\mu}\right)^n\cdot\frac{1}{k^n} = \frac{k^k}{k!}\sum_{n=k}^{\infty}\left(\frac{\lambda}{k\mu}\right)^n$$

and the last series converges if and only if $\dfrac{\lambda}{k\mu} < 1$; that is $\lambda < k\mu$.

Example

An auto licensing office has six agents. Customers arrive according to a Poisson distribution on average at two per minute and are served according to an exponential distribution. The average time that an agent spends with a customer is 10 min. We analyze the dynamics of the office.

We have

$$\lambda = \frac{1}{2} = .5, \quad \mu = \frac{1}{10} = .1, \quad k = 6 \quad \text{so} \quad \frac{\lambda}{\mu} = 5 \quad \text{and} \quad \rho = \frac{\lambda}{k\mu} = \frac{5}{6}.$$

Since $\rho < 1$, the system has a unique equilibrium state.

The probability that no customer is in the office is

$$\pi(0) = \pi(0) = \left[\sum_{n=0}^{k-1} \frac{(k\rho)^n}{n!} + \frac{k^k}{k!} \cdot \frac{\rho^k}{(1-\rho)}\right]^{-1} = \left[\sum_{n=0}^{5} \frac{\left(6 \cdot \frac{5}{6}\right)^n}{n!} + \frac{6^6}{6!} \cdot \frac{\left(\frac{5}{6}\right)^6}{\left(1 - \frac{5}{6}\right)}\right]^{-1}$$

$$= \left[\frac{1097}{12} + \frac{3125}{24}\right]^{-1} = \left[\frac{1773}{8}\right]^{-1} = \frac{8}{1773} \approx .0045.$$

The probability an entering customer will have to wait in the queue is

$$\pi(0)\frac{k^k}{k!}\frac{\rho^k}{1-\rho} = \frac{8}{1773} \cdot \frac{6^6}{6!} \cdot \frac{\left(\frac{5}{6}\right)^6}{1 - \frac{5}{6}} = \frac{3125}{5319} \approx .588.$$

The expected length of the queue is

$$L_Q = \pi(0)\frac{k^k}{k!}\rho^k \frac{\rho}{(1-\rho)^2} = \frac{8}{1773} \cdot \frac{6^6}{6!} \cdot \left(\frac{5}{6}\right)^6 \cdot \frac{\left(\frac{5}{6}\right)}{\left(1 - \frac{5}{6}\right)^2} = \frac{15{,}625}{5{,}319} \approx 2.94,$$

the expected waiting time in the queue is

$$W_Q = \frac{L_Q}{\lambda} = \frac{\frac{15{,}625}{5{,}319}}{\frac{1}{2}} = \frac{31{,}250}{5{,}329} \approx 5.88,$$

the expected total time in the system is

$$W = \frac{L_Q}{\lambda} + \frac{1}{\mu} = \frac{31250}{5329} + \frac{1}{\frac{1}{10}} = \frac{84,440}{5,319} \approx 15.88,$$

and the average number of people in the office is

$$L = \lambda W = \frac{1}{2} \cdot \frac{84,440}{5,319} = \frac{42,220}{5,319} \approx 7.94.$$

M/M/c/K Queue: c Servers (Size of the Queue Is Limited to K)

The analysis of this model is an amalgam of the model with one server and a limited capacity for the queue and the model with c servers and no limit on the capacity of the queue. We will present only the results and some examples.

The transition rate diagram for this model is shown in Figure 5.7.

We assume that $K \geq c$. To determine the infinitesimal generator, we use the first $(c + 1) \times (c + 1)$ block of the generator for the M/M/c model, which is

$$A = \begin{pmatrix} -\lambda & \lambda & 0 & 0 & & \cdots & 0 \\ \mu & -(\mu+\lambda) & \lambda & 0 & & \cdots & 0 \\ 0 & 2\mu & -(2\mu+\lambda) & \lambda & 0 & & 0 \\ \vdots & \vdots & \vdots & \vdots & \ddots & & \vdots \\ 0 & & & \cdots & 0 & c\mu & -(c\mu+\lambda) \quad \lambda \end{pmatrix}.$$

For the remaining states, similar to the M/M/1/K model, the lower right part of the matrix is

$$B = \begin{pmatrix} c\mu & -(c\mu+\lambda) & \lambda & 0 & \cdots & 0 \\ 0 & c\mu & -(c\mu+\lambda) & \lambda & \cdots & 0 \\ \vdots & & & \vdots & & \\ 0 & \cdots & 0 & c\mu & & -(c\mu) \end{pmatrix}.$$

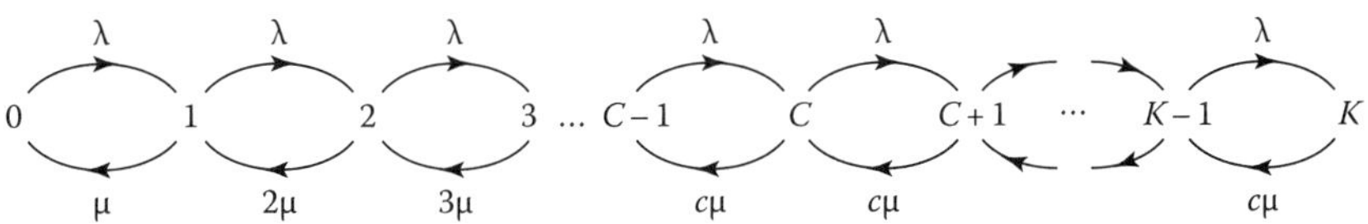

FIGURE 5.7
Transition rate diagram for M/M/c/K model.

The infinitesimal generator is the $(K + 1) \times (K + 1)$ matrix

$$G = \begin{pmatrix} A & 0 \\ 0 & B \end{pmatrix}.$$

The components of the steady state are

$$\pi_0 = \left[\sum_{n=0}^{c} \left(\frac{\lambda}{\mu}\right)^n \frac{1}{n!} + \frac{1}{c!} \sum_{n=c+1}^{K} \left(\frac{\lambda}{\mu}\right)^n \frac{1}{c^{n-c}} \right]^{-1};$$

$$\pi_j = \frac{\left(\frac{\lambda}{\mu}\right)^j}{j!} \pi_0, \quad j = 1, \ldots, c;$$

$$\pi_j = \frac{\left(\frac{\lambda}{\mu}\right)^j}{c^{j-c} c!} \pi_0, \quad j = c+1, \ldots, K.$$

The average number of customers in the system is

$$\frac{\lambda}{\mu} + \frac{\rho (c\rho)^c}{(1-\rho)^2 c!} \pi_0, \quad \text{where } \rho = \frac{\lambda}{c\mu}.$$

Example

Suppose we have a system in which there are two servers and the system has the capacity of three customers. Suppose that customers arrive according to a Poisson process with rate $\lambda = 4$ and service time of a customer is exponentially distributed at the rate $\mu = 8$. We compute the equilibrium state of the system.

The infinitesimal generator is

$$G = \begin{pmatrix} -4 & 4 & 0 & 0 \\ 8 & -12 & 4 & 0 \\ 0 & 16 & -20 & 4 \\ 0 & 0 & 16 & -16 \end{pmatrix}.$$

Computing the equilibrium state, we have

$$\pi_0 = \left[\sum_{n=0}^{c}\left(\frac{\lambda}{\mu}\right)^n \frac{1}{n!} + \frac{1}{c!}\sum_{n=c+1}^{K}\left(\frac{\lambda}{\mu}\right)^n \frac{1}{c^{n-c}}\right]^{-1}$$

$$= \left[\sum_{n=0}^{2}\left(\frac{4}{8}\right)^n \frac{1}{n!} + \frac{1}{2!}\sum_{n=3}^{3}\left(\frac{4}{8}\right)^n \frac{1}{2^{n-2}}\right]^{-1}$$

$$= \left[1+\left(\frac{4}{8}\right)+\left(\frac{4}{8}\right)^2\frac{1}{2!}+\frac{1}{2!}\left(\frac{4}{8}\right)^3\frac{1}{2^1}\right]^{-1} = \left(\frac{53}{32}\right)^{-1} = \frac{32}{53};$$

$$\pi_1 = \left(\frac{\lambda}{\mu}\right)\pi_0 = \frac{16}{53};\quad \pi_2 = \left(\frac{\lambda}{\mu}\right)^2\frac{1}{2!}\pi_0 = \frac{4}{53};\quad \pi_3 = \frac{\left(\frac{\lambda}{\mu}\right)^3}{2^{3-2}2!}\pi_0 = \frac{1}{53}.$$

Thus,

$$\pi = \left(\frac{32}{53},\frac{16}{53},\frac{4}{53},\frac{1}{53}\right)$$

is the equilibrium state. We note that

$$\pi G = \left(\frac{32}{53},\frac{16}{53},\frac{4}{53},\frac{1}{53}\right)\begin{pmatrix} -4 & 4 & 0 & 0 \\ 8 & -12 & 4 & 0 \\ 0 & 16 & -20 & 4 \\ 0 & 0 & 16 & -16 \end{pmatrix} = (0,0,0,0)$$

as required.

We solve the same problem using detailed balance equations. We have

$$\pi_0 G(0,1) = \pi_1 G(1,0),\quad \text{so } 4\pi_0 = 8\pi_1 \text{ and } \pi_1 = \frac{1}{2}\pi_0;$$

$$\pi_1 G(1,2) = \pi_2 G(2,1),\quad \text{so } 4\pi_1 = 16\pi_2 \text{ and } \pi_2 = \frac{1}{4}\pi_1 = \frac{1}{8}\pi_0;$$

$$\pi_2 G(2,3) = \pi_3 G(3,2),\quad \text{so } 4\pi_2 = 16\pi_3 \text{ and } \pi_3 = \frac{1}{4}\pi_2 = \frac{1}{32}\pi_0.$$

Then

$$1 = \pi_0 + \pi_1 + \pi_2 + \pi_3 = \pi_0\left(1+\frac{1}{2}+\frac{1}{8}+\frac{1}{32}\right) = \pi_0\left(\frac{53}{32}\right)$$

TABLE 5.2

Formulas for the M/M/K and M/M/c/K Models

	M/M/K	M/M/c/K
P_0	$\dfrac{1}{\sum_{n=0}^{k-1}\dfrac{\left(\frac{\lambda}{\mu}\right)^n}{n!}+\dfrac{\left(\frac{\lambda}{\mu}\right)^k}{k!\left[1-\lambda/(k\mu)\right]}}$	$\dfrac{1}{\sum_{n=0}^{c}\dfrac{\left(\frac{\lambda}{\mu}\right)^n}{n!}+\dfrac{\left(\frac{\lambda}{\mu}\right)^c}{c!}\sum_{n=c+1}^{K}\left(\frac{\lambda}{c\mu}\right)^{n-c}}$
P_n	$\begin{cases}\dfrac{\left(\frac{\lambda}{\mu}\right)^n}{n!}P_0, & n=1,\ldots,k\\ \dfrac{\left(\frac{\lambda}{\mu}\right)^n}{k!k^{n-k}}P_0, & n=k+1,\ldots\end{cases}$	$\begin{cases}\dfrac{\left(\frac{\lambda}{\mu}\right)^n}{n!}P_0, & n=1,\ldots,c\\ \dfrac{\left(\frac{\lambda}{\mu}\right)^n}{k!k^{n-k}}P_0, & n=c+1,\ldots K\end{cases}$
L	$P_0\dfrac{\left(\frac{\lambda}{\mu}\right)^{k+1}}{k!\left(1-\frac{\lambda}{\mu}\right)^2}+\dfrac{\lambda}{\mu}$	$\sum_{n=0}^{c-1}nP_n+L_q+c\left(1-\sum_{n=0}^{c-1}P_n\right)$
L_q	$P_0\dfrac{\left(\frac{\lambda}{\mu}\right)^{k+1}}{k!\left(1-\frac{\lambda}{\mu}\right)^2}$	$P_0\dfrac{\left(\frac{\lambda}{\mu}\right)^c\left(\frac{\lambda}{\mu}\right)\left[1-\left(\frac{\lambda}{\mu}\right)^{K-c}-(K-c)\left(\frac{\lambda}{\mu}\right)^{K-c}\left(1-\frac{\lambda}{\mu}\right)\right]}{c!\left(1-\frac{\lambda}{\mu}\right)^2}$
W	$\dfrac{L_q}{\lambda}+\dfrac{1}{\mu}$	$\dfrac{L}{\lambda(1-P_k)}$
W_q	$\dfrac{L_q}{\lambda}$	$\dfrac{L_q}{\lambda(1-P_k)}$

and so

$$\pi_0=\frac{32}{53},\quad \pi_1=\frac{16}{53},\quad \pi_2=\frac{4}{53},\quad \pi_3=\frac{1}{53},$$

as before.

Table 5.2 gives a summary of the formulas for the M/M/K and M/M/c/K queuing models. The quantities are

P_0 = probability the system is in state 0 at equilibrium

P_n = probability the system is in state n at equilibrium

L = average number of customers in the system

L_q = average number of customers in the queue

W = average waiting time in the system

W_q = average waiting time in the queue

M/M/∞ Model

We repeat the analysis of the M/M/K case except that $\mu_k = k\mu$ for all k. The infinitesimal generator is

$$\begin{pmatrix} -\lambda & \lambda & 0 & 0 & 0 & 0 & 0 & 0 & 0 & 0 & \cdots \\ 0 & \mu & -(\mu+\lambda) & \lambda & 0 & 0 & 0 & 0 & 0 & 0 & \cdots \\ 0 & 0 & 2\mu & -(2\mu+\lambda) & \lambda & 0 & 0 & 0 & 0 & 0 & \cdots \\ 0 & 0 & 0 & 3\mu & -(3\mu+\lambda) & \lambda & 0 & 0 & 0 & 0 & \cdots \\ \vdots & \vdots & \vdots & & & & & & & & \end{pmatrix}.$$

The detailed balance equations are

$$\lambda\pi_0 = \mu\pi_1,$$

$$\lambda\pi_n = (n+1)\,\mu\pi_{n+1}$$

so

$$\pi_{n+1} = \frac{\lambda}{(n+1)\,\mu}\pi_n = \left(\frac{\lambda}{\mu}\right)^{n+1} \cdot \frac{1}{(n+1)!}\pi_0.$$

Now

$$1 = \sum_{n=0}^{\infty} \pi_n = \pi_0 \left[1 + \cdots + \left(\frac{\lambda}{\mu}\right)^n \cdot \frac{1}{n!} + \cdots\right] = \pi_0 e^{\left(\frac{\lambda}{\mu}\right)}$$

so

$$\pi_0 = e^{-\left(\frac{\lambda}{\mu}\right)} \quad \text{and} \quad \pi_n = \frac{\left(\frac{\lambda}{\mu}\right)^n}{n!}\pi_0 = e^{-\left(\frac{\lambda}{\mu}\right)} \frac{\left(\frac{\lambda}{\mu}\right)^n}{n!}.$$

Also, if $p_{0j}(t)$ is the probability that the system is in state j at time t, having begun in state 0, then

$$p_{0j}(t) = \exp\left[-\left(\frac{\lambda}{\mu}\right)\left(1-e^{-\mu t}\right)\right]\frac{\left[\left(\frac{\lambda}{\mu}\right)\left(1-e^{-\mu t}\right)\right]^j}{j!}.$$

The derivation of this formula is beyond the scope of the text.

The average number of customers in the system at equilibrium is

$$E[N] = L = \sum_{n=0}^{\infty} n\pi_n = \sum_{n=1}^{\infty} ne^{-\lambda/\mu}\frac{(\lambda/\mu)^n}{n!} = e^{-\lambda/\mu} n\sum_{n=1}^{\infty}\frac{(\lambda/\mu)^n}{n!}$$

$$= e^{-\lambda/\mu}(\lambda/\mu)\sum_{n=1}^{\infty}\frac{(\lambda/\mu)^{n-1}}{(n-1)!} = e^{-\lambda/\mu}(\lambda/\mu)\sum_{n=0}^{\infty}\frac{(\lambda/\mu)^n}{n!}$$

$$= e^{-\lambda/\mu}(\lambda/\mu)e^{\lambda/\mu} = \frac{\lambda}{\mu}.$$

Exercises

5.1 Give the infinitesimal generator for a Poisson process with $\lambda = 2$.

5.2 For the following infinitesimal generators,

(a) Find the transition matrix for the embedded Markov chain

(b) Find the equilibrium state for the embedded Markov chain

(c) Find the equilibrium state for the continuous Markov chain

(d) Find the eigenvalues and eigenvectors

(e) Give the solution for the Kolmogorov forward equations:

$$\text{(i)}\quad G = \begin{pmatrix} -4 & 1 & 3 \\ 2 & -6 & 4 \\ 3 & 2 & -5 \end{pmatrix}$$

$$\text{(ii)}\quad G = \begin{pmatrix} -2 & 1 & 1 \\ 1 & -5 & 4 \\ 2 & 1 & -3 \end{pmatrix}$$

$$\text{(iii)}\quad G=\begin{pmatrix} -8 & 3 & 1 & 4 \\ 2 & -4 & 1 & 1 \\ 3 & 1 & -6 & 2 \\ 2 & 1 & 2 & 5 \end{pmatrix}.$$

5.3 Consider a reflecting walk on the nonnegative integers, with

$$P_{0,1}=1;\quad P_{i,i+1}=\frac{1}{3},\quad P_{i,i-1}=\frac{2}{3},\quad i\geq 1.$$

Let $\{\lambda_0, \lambda_1, \lambda_2, \ldots\}$ be a convergent sequence of positive numbers. Construct a continuous Markov process by having the transition rate at state i exponentially distributed with parameter λ_i.

(a) Find the infinitesimal generator of the Markov chain.

(b) Find the equilibrium distribution for the Markov chain.

5.4 Suppose we have a generator that fails on the average of once every 8 months, with exponential distribution. Suppose the average repair time is 1/4 of a month, exponentially distributed.

(a) Find the infinitesimal generator of the Markov process described by this situation, where state 0 is that the generator is functioning and state 1 is that the generator is in for repair.

(b) If the generator is a function at time $t = 0$, what is the probability it will be in for repairs 15 months later?

(c) What is the steady state of this Markov chain?

5.5 There are 3 checkout counters at a store. Customers arrive according to a Poisson distribution with $\lambda = 4$ and are served at an exponential rate with $\mu = 6$.

(a) What is meant by the equilibrium state of the system? What is the equilibrium state of this system?

(b) What is the probability that an arriving customer will have to wait in line as opposed to going to checkout immediately?

(c) What is the expected number of customers in line when the system is in equilibrium?

(d) What is the expected waiting time in line when the system is in equilibrium?

(e) What is the expected number of customers in the system when the system is in equilibrium?

5.6 Repeat problem 5 with $\lambda = 4$ and $\mu = 8$.

5.7 The local NPR station is doing a telethon. There is only one person actually receiving calls, but up to four callers can be put on hold.

Calls come in according to a Poisson distribution with $\lambda = 4$ and callers are serviced according to an exponential distribution with $\mu = 3$.

(a) Find the equilibrium distribution of the calls in the system.

(b) Find the probability that there are no callers at a given time.

(c) Find the expected number of callers in the system at a given time.

(d) Find the rate at which callers cannot get through.

(e) Find the rate at which callers cannot get through if up to six callers could be put on hold.

5.8 A bank has a drive-in service of one window. Customers arrive according to a Poisson distribution at an average of one every 6 min. Customers are serviced at an exponential rate with a mean of 5 min. An arriving customer will not enter the system if there are three cars in the line, including the customer being served.

(a) Find the equilibrium distribution of the customers in the system.

(b) Find the probability that there is exactly one customer in the system.

(c) Find the expected number of customers in the system.

(d) Find the rate at which potential customers do not enter the system.

5.9 We have a rat in a house that is shown in Figure 5.8.

The rat stays in room k an amount of time that has an exponential distribution with parameter λ_k, and if there are n doors in the room it is in, it then exits each door with probability $\frac{1}{n}$.

(a) If $\lambda_1 = 3, \lambda_2 = 1, \lambda_3 = 2, \lambda_4 = 10$, find the infinitesimal generator.

(b) Find the equilibrium state.

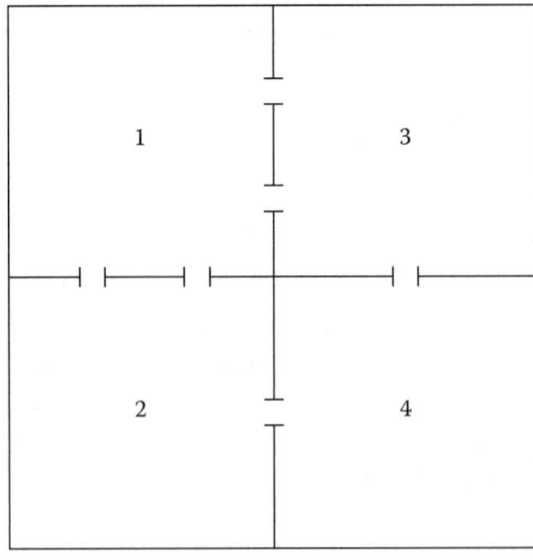

FIGURE 5.8
House for a mouse.

(c) Find $P_{ij}(t)$, $1 \le i, j \le 4$.

(d) Find the equilibrium state if $\lambda_4 = 1$ and all other parameters remain the same.

5.10 Prove that for P, a transition matrix, and λ, the rate of change for the related continuous-time Markov process,

(a) $\lim\limits_{t \downarrow 0} \dfrac{1 - P_{ii}(t)}{t} = \lambda(i)$.

(b) $\lim\limits_{t \downarrow 0} \dfrac{1 - P_{ii}(t)}{t} = \lambda(i) P_{ij},\ i \ne j$.

5.11 In this problem, we consider an M/M/1/1 queue. That is, we have one server and a customer will leave if she cannot be served immediately. Suppose customers arrive according to a Poisson distribution with $\lambda = 1$ and service is performed according to an exponential distribution with $\mu = 1$.

(a) Give the infinitesimal distribution of the process.

(b) Find the equilibrium state.

(c) What is the rate at which potential customers are turned away?

5.12 This problem refers to an M/M/1 queue. Suppose that customers arrive at the rate λ and are processed at the rate μ with $\lambda < \mu$.

(a) Show that the time waiting in the queue has the cumulative distribution function

$$F(t) = 1 - \left(\frac{\lambda}{\mu}\right) e^{-(\mu - \lambda)}.$$

(b) If customers arrive at the exponential rate of 2 per hour and are processed at the exponential rate of 3 per hour, what is the probability that a customer will spend less than 15 min in the queue?

5.13 Consider the case where we have M machines and s service people, where $s < M$. Each machine fails at an exponential rate of λ, and each service person repairs a machine at an exponential rate of μ.

Let n be the number of machines requiring service. In this case, we say the system is in state n.

If there are M machines and there are n machines requiring service, then there are $(M - n)$ machines that are working, so the failure rate when the system is in state n is λ_n where

$$\lambda_n = (M - n)\lambda.$$

If there are n machines requiring service and $n \le s$, then the repair rate is μ_n where $\mu_n = n\mu$. If $n > s$, then the repair rate is $\mu_n = s\mu$. Thus,

$$\mu_n = \begin{cases} n\mu & \text{if } n \le s \\ s\mu & \text{if } n > s \end{cases}.$$

Let $\pi = (\pi_0, \pi_1, \ldots, \pi_M)$ denote the equilibrium state.

(a) Use the detailed balance equations to show

$$\pi_1 = \frac{M\lambda}{\mu}\pi_0.$$

(b) Show that

$$\pi_2 = \frac{(M-1)\lambda}{2\mu}\pi_1 = \frac{M(M-1)\lambda^2}{2\mu^2}\pi_0$$

and

$$\pi_3 = \frac{(M-2)\lambda}{3\mu}\pi_2 = \frac{M(M-1)(M-2)\lambda^3}{3\cdot 2\mu^3}\pi_0 = \binom{M}{3}\left(\frac{\lambda}{\mu}\right)^3\pi_0.$$

(c) Show that if $k < s$, then

$$\pi_{k+1} = \frac{(M-k)\lambda}{(k+1)\mu}\pi_k = \binom{M}{k}\left(\frac{\lambda}{\mu}\right)^k\pi_0.$$

(d) Suppose that $M = 10$ and $s = 3$. Then

$$\lambda_0\lambda_1\ldots\lambda_6 = (10)(10-1)(10-2)(10-3)(10-4)(10-5)(10-6)\lambda^7$$

$$\mu_1\mu_2\ldots\mu_7 = (\mu)(2\mu)(3\mu)(3\mu)(3\mu)(3\mu)(3\mu) = 3!3^{7-3}\mu^7.$$

Now

$$\pi_7 = \frac{\lambda_0\,\lambda_1\ldots\lambda_6}{\mu_1\,\mu_2\ldots\mu_7}\pi_0 = \frac{(10)(10-1)\ldots(10-6)}{3!3^{7-3}}\left(\frac{\lambda}{\mu}\right)^7.$$

Show that

$$\frac{(10)(10-1)\dots(10-6)}{3!3^{7-3}} = \binom{10}{7}\frac{7!}{3!3^{7-3}}$$

and thus

$$\pi_7 = \binom{10}{7}\frac{7!}{3!3^{7-3}}\left(\frac{\lambda}{\mu}\right)^7 \pi_0.$$

(e) Use the ideas of part (d) to show that if there are M machines and s repairmen, then for $s \le n \le M$,

$$\pi_n = \binom{M}{n}\frac{n!}{s!s^{n-s}}\left(\frac{\lambda}{\mu}\right)^n \pi_0.$$

5.14 We have 3 machines and two repairmen. The machines work for an exponentially distributed time before breaking down with a mean of 8, and a repairman can repair a machine according to an exponentially distributed time with mean 1. Let n be the number of machines requiring service. In this case, we say the system is in state n.

(a) Find the equilibrium distribution of the system.

(b) If it requires at least 2 machines to keep the facility running, what is the probability the facility will be down?

(c) What is the expected number of machines in service?

5.15 (a) Find the transition matrix for the SIS model when $N = 3$, $\beta = \frac{1}{3}$, and $\gamma = \frac{1}{2}$.

(b) Use a CAS and your answer to part (a) to find the equilibrium state for the SIS model when $N = 3$, $\beta = \frac{1}{3}$, and $\gamma = \frac{1}{2}$.

6

Reversible Markov Chains

In this chapter, we study a class of Markov chains that are *time reversible* or *reversible*. Many important Markov chains fall into this category, and sometimes, additional insights can be obtained, or complex problems can be simplified by considering the reverse property.

A metaphor that is often used to describe a reversible Markov chain is a movie. Most movies one can tell if it is being run forward or backward, but it is possible to conceive of examples where this is not the case. For a movie of a person walking, the direction of the movie is obvious, but this is not the case of a recording of a person jumping up and down.

An intuitive description (which will need to be made more precise) that reflects the movie metaphor is that a discrete-time Markov process $\{X_n\}$ is reversible if, when in the equilibrium state, the distribution of $(X_0, X_1, \ldots, X_n)$ is the same as $(X_n, X_{n-1}, \ldots, X_0)$ for all positive integers n.

The property of being reversible is dependent on there being a single equilibrium distribution, so in this chapter all the Markov chains that we discuss will be assumed to be irreducible and positive recurrent. The state space is countable (as always, this includes finite), and the time may be discrete or continuous. Our initial discussion will be for the discrete case.

Theorem 6.1

Let $\{X_n\}$ be a positive recurrent, irreducible Markov chain with equilibrium distribution π. If the initial distribution is π, then the reversed process satisfies the Markov property. That is,

$$P(X_k = i \mid X_{k+1} = j, X_{k+2} = i_2, X_{k+3} = i_3, \ldots, X_{k+n} = i_n) = P(X_k = i \mid X_{k+1} = j)$$

for every positive integer k.

NOTE: This does not say that $\{X_n\}$ is reversible. In fact, in order to have time homogeneity for the reversed process, more is required.

Proof

Fix a positive integer k. We will show

$$P(X_k = i \mid X_{k+1} = j, X_{k+2} = i_2, X_{k+3} = i_3, \ldots, X_{k+n} = i_n) = P(X_k = i \mid X_{k+1} = j).$$

Now,

$$P(X_k = i \mid X_{k+1} = j, X_{k+2} = i_2, \ldots, X_{k+n} = i_n) = \frac{P(X_k = i, X_{k+1} = j, \ldots, X_{k+n} = i_n)}{P(X_{k+1} = j, X_{k+2} = i_2, \ldots, X_{k+n} = i_n)}$$

$$= \frac{\pi(i) P_{ij} P_{j,i_2} \cdots P_{i_{n-1},i_n}}{\pi(j) P_{j,i_2} \cdots P_{i_{n-1},i_n}} = \frac{\pi(i) P_{ij}}{\pi(j)}.$$

But

$$P(X_k = i \mid X_{k+1} = j) = \frac{P(X_k = i, X_{k+1} = j)}{P(X_{k+1} = j)} = \frac{\pi(i) P_{ij}}{\pi(j)},$$

so the Markov property is satisfied.

In order for the Markov chain to be reversible, we must have

$$P(X_{k+1} = i \mid X_k = j) = P(X_k = j \mid X_{k+1} = i).$$

But

$$P(X_{k+1} = i \mid X_k = j) = P_{ji}$$

and under the hypothesis of the theorem,

$$P(X_k = i \mid X_{k+1} = j) = \frac{\pi(i) P_{ij}}{\pi(j)}.$$

Thus, we have the following crucial result:

Corollary

A positive recurrent discrete-time Markov chain with transition matrix P is reversible if there is probability distribution π such that all states i and j

$$\pi(i) P_{ij} = \pi(j) P_{ji}. \tag{1}$$

We will show in the next theorem that π in equation (1) is an equilibrium distribution.

While there are different equivalent ways of characterizing a reversible Markov process, the equations given by (1) provide the most concise way to show that a positive recurrent discrete-time Markov chain with transition matrix P is reversible.

The equations given by (1) are the detailed balance equations that we dealt with in the previous chapter and are particularly important in reversible Markov chains.

A physical interpretation of these equations can be thought of in terms of a *probability flux*. They say that when the process is in the equilibrium state the probability of passing from state i to state j is equal to the probability of passing from state j to state i.

Theorem 6.2

A probability distribution π that satisfies $\pi(i)P_{ij} = \pi(j)P_{ji}$ is an equilibrium state.

Proof

Suppose for each state i

$$P(X_0 = i) = \pi(i)$$

where π is the probability distribution π that satisfies $\pi(i)P_{ij} = \pi(j)P_{ji}$.

For each i, j we have

$$\pi(i)P(X_1 = j \mid X_0 = i) = \pi(i)P_{ij}.$$

By the hypothesis,

$$\pi(i)P_{ij} = \pi(j)P_{ji}$$

and

$$\pi(j)P_{ji} = \pi(j)P(X_1 = i \mid X_0 = j)$$

so

$$\begin{aligned} P(X_1 = j) &= \sum_i \pi(i)P(X_1 = j \mid X_0 = i) \\ &= \sum_i \pi(j)P(X_1 = i \mid X_0 = j) \\ &= \pi(j)\sum_i P(X_1 = i \mid X_0 = j) = \pi(j) = P(X_0 = j). \end{aligned}$$

It follows by time homogeneity that for each state j and each time n, we have $P(X_{n+1} = j) = P(X_n = j)$ and thus $P(X_n = j) = P(X_0 = j)$.

Corollary

If the process begins in the steady-state π, then for all times n, we have

$$P(X_n = i, X_{n+1} = j) = P(X_{n+1} = i, X_n = j).$$

We have $\pi(i)P_{ij}$ is the probability of seeing a transition from state i to state j, and $\pi(j)P_{ji}$ is the probability of seeing a transition from state j to state i. Another way to think of a time-reversible Markov chain is a positive recurrent Markov chain is time reversible if for all states i and j, the rate at which the process changes from state i to state j is the same as the rate at which the process changes from state j to state i.

The two conditions,

1. The distribution of $(X_0, X_1, \ldots, X_n)$ is the same as $(X_n, X_{n-1}, \ldots, X_0)$ for all positive integers n
2. $\pi(i)P_{ij} = \pi(j)P_{ji}$ for all i, j

are equivalent as shown in Theorem 6.3, but there is an underlying fact that should be emphasized. Namely, that the distribution π is the unique equilibrium distribution and the rate of change between the states being equal assumes that the process is in equilibrium.

Theorem 6.3

Let $\{X_n, n \geq 0\}$ be an irreducible, positive recurrent, discrete-time Markov process with transition matrix P and unique equilibrium distribution π. The following conditions are equivalent:

(a) For all states i and j, $\pi(i)P_{ij} = \pi(j)P_{ji}$.
(b) If the initial distribution is π, then $(X_0, X_1, \ldots, X_n)$ has the same distribution as $(X_n, X_{n-1}, \ldots, X_0)$ for all n.

Either condition can be taken as the definition of a reversible Markov chain.

Condition (a) emphasizes that when in equilibrium, the rate at which the process changes from state i to state j is the same as the rate at which the process changes from state j to state i, and condition (b) emphasizes the process is the same running forward or backward.

Proof

(a) $\Rightarrow$ (b) We show $(X_0, X_1, \ldots, X_n)$ has the same distribution as $(X_n, X_{n-1}, \ldots, X_0)$ for all n.

Showing that (X_0, X_1) has the same distribution as (X_1, X_0) in the equilibrium state is the same as

$$P(X_0 = i, X_1 = j) = P(X_1 = i, X_0 = j) \quad \text{for all states } i \text{ and } j.$$

But $P(X_0 = i, X_1 = j)$ is the probability that the process begins in state i and transitions to state j on the first step. Since we are assuming the initial distribution is π, then $\pi(i)$ is the probability that the process begins in state i.

Also

$$P_{ij} \text{ is the probability the process transitions from } i \text{ to } j$$

so

$$P(X_0 = i, X_1 = j) = \pi(i)P_{ij}.$$

Likewise,

$$P(X_1 = i, X_0 = j) = \pi(j)P_{ji},$$

and since we are assuming $\pi(i)P_{ij} = \pi(j)P_{ji}$, then (X_0, X_1) has the same distribution as (X_1, X_0).

We next show that (X_0, X_1, X_2) has the same distribution as (X_2, X_1, X_0).

To show (X_0, X_1, X_2) has the same distribution as (X_2, X_1, X_0) is the same as showing

$$P(X_0 = i, X_1 = j, X_2 = k) = P(X_2 = i, X_1 = j, X_0 = k)$$

for all states i, j, and k.

Now, $P(X_0 = i, X_1 = j, X_2 = k)$ is the probability that the process begins in state i, transitions to state j on the first step, and transitions to state k on the second step. So

$$P(X_0 = i, X_1 = j, X_2 = k) = \left[\pi(i)P_{ij}\right]P_{jk} = \left[\pi(j)P_{ji}\right]P_{jk} = \left[\pi(j)P_{jk}\right]P_{ji}$$
$$= \left[\pi(k)P_{kj}\right]P_{ji}.$$

But $\left[\pi(k)P_{kj}\right]P_{ji}$ is the probability that the process begins in state k, transitions to state j on the first step, and transitions to state i on the second step, which is $P(X_0 = k, X_1 = j, X_2 = i) = P(X_2 = i, X_1 = j, X_0 = k)$ so

$$P(X_0 = i, X_1 = j, X_2 = k) = P(X_2 = i, X_1 = j, X_0 = k),$$

and thus (X_0, X_1, X_2) has the same distribution as (X_2, X_1, X_0).

This pattern works to show that for n a positive integer,

$$P(X_0 = i_0, X_1 = i_1, \ldots, X_n = i_n) = \left[\pi(i_0)P_{i_0,i_1}\right]P_{i_1,i_2}P_{i_2,i_3}\cdots P_{i_{n-1},i_n}$$

and

$$\begin{aligned}
\left[\pi(i_0)P_{i_0,i_1}\right]P_{i_1,i_2}P_{i_2,i_3}\cdots P_{i_{n-1},i_n} &= \left[\pi(i_1)P_{i_1,i_0}\right]P_{i_1,i_2}P_{i_2,i_3}\cdots P_{i_{n-1},i_n} \\
&= P_{i_1,i_0}\left[\pi(i_1)P_{i_1,i_2}\right]P_{i_2,i_3}\cdots P_{i_{n-1},i_n} \\
&= P_{i_1,i_0}\left[\pi(i_2)P_{i_2,i_1}\right]P_{i_2,i_3}\cdots P_{i_{n-1},i_n} \\
&= P_{i_1,i_0}P_{i_2,i_1}\left[\pi(i_2)P_{i_2,i_3}\right]P_{i_3,i_4}\cdots P_{i_{n-1},i_n} \\
&= P_{i_1,i_0}P_{i_2,i_1}P_{i_2,i_3}\left[\pi(i_3)P_{i_3,i_4}\right]P_{i_4,i_5}\cdots P_{i_{n-1},i_n} \\
&\ \ \vdots \\
P_{i_1,i_0}P_{i_2,i_1}\cdots\left[\pi(i_{n-1})P_{i_{n-1},i_n}\right] &= P_{i_1,i_0}P_{i_2,i_1}\cdots\left[\pi(i_n)P_{i_n,i_{n-1}}\right] \\
&= P_{i_1,i_0}P_{i_2,i_1}\cdots P_{i_n,i_{n-1}}\pi(i_n) = \pi(i_n)P_{i_n,i_{n-1}}\cdots P_{i_2,i_1}P_{i_1,i_0} \\
&= P(X_0 = i_n, X_1 = i_{n-1}, \ldots, X_n = i_0) \\
&= P(X_n = i_0, X_{n-1} = i_1, \ldots, X_0 = i_n).
\end{aligned}$$

Thus, we have

$$P(X_0 = i_0, X_1 = i_1, \ldots, X_n = i_n) = P(X_n = i_0, X_{n-1} = i_1, \ldots, X_0 = i_n),$$

and so the distribution of $(X_0, X_1, \ldots, X_n)$ is the same as the distribution of $(X_n, X_{n-1}, \ldots, X_0)$ for all n.

(b) ⇒ (a)

Taking $n = 1$, we have that (X_0, X_1) has the same distribution as (X_1, X_0), so for all states i and j,

$$P(X_0 = i, X_1 = j) = P(X_i = i, X_0 = j)$$

but

$$P(X_0 = i, X_1 = j) = \pi(i)P_{ij}$$

and

$$P(X_i = i, X_0 = j) = P(X_0 = j, X_0 = i) = \pi(j)P_{ji}$$

and so (a) holds.

The following corollary is the most important result of the chapter.

Corollary

An irreducible, positive recurrent Markov chain is reversible if and only if for the equilibrium state π, we have

$$\pi(i)P_{ij} = \pi(j)P_{ji} \quad \text{for all } i, j.$$

Example

In a previous example, we showed that (.75, .25) was the equilibrium state for the Markov chain whose transition matrix is

$$P = \begin{pmatrix} .9 & .1 \\ .3 & .7 \end{pmatrix}.$$

We show that the Markov chain is reversible by showing $\pi(1)P_{12} = \pi(2)P_{21}$. We have

$$\pi(1) = .75,\ \pi(2) = .25; \quad P_{12} = .1,\ P_{21} = .3$$

so

$$\pi(1)P_{12} = .075 \quad \text{and} \quad \pi(2)P_{21} = .075,$$

and thus, the Markov chain is reversible.

In Exercise 6.5, we show that the two state Markov chains whose transition matrix is

$$P = \begin{pmatrix} \alpha & 1-\alpha \\ 1-\beta & \beta \end{pmatrix}, \quad 0 < \alpha, \beta < 1$$

is reversible.

Theorem 6.4 (Kolmogorov's Cycle Criterion, Discrete-Time Version)

A discrete-time Markov chain is reversible if and only if the product of the transition probabilities along any loop is equal to the product of the transition probabilities in the reverse order. That is, for any states $i, i_1, i_2, \ldots, i_{k-1}, i_k, j$, we have

$$p(i,i_1)p(i_1,i_2)\cdots p(i_{k-1},i_k)p(i_k,j) = p(j,i_k)p(i_k,i_{k-1})\cdots p(i_2,i_1)p(i_1,i).$$

Proof

Suppose that for any states $i, i_1, i_2, \ldots, i_{k-1}, i_k, j$, we have

$$p(i,i_1)p(i_1,i_2)\cdots p(i_k,j)p(j,i) = p(i,j)p(j,i_k)\cdots p(i_2,i_1)p(i_1,i).$$

Consider the case for $k = 2$. We have

$$p(i,i_1)p(i_1,i_2)p(i_2,j) = p(j,i_2)p(i_2,i_1)p(i_1,i).$$

Now,

$$\sum_{i_1} p(i,i_1)p(i_1,i_2) = p^2(i,i_2) \quad \text{and} \quad \sum_{i_2} p^2(i,i_2)p(i_2,j) = p^3(i,j)$$

so

$$\sum_{i_2}\left(\sum_{i_1} p(i,i_1)p(i_1,i_2)p(i_2,j)p(j,i)\right) = p^3(i,j)p(j,i).$$

Similarly,

$$\sum_{i_2}\left(\sum_{i_1} p(i,j)p(j,i_2)p(i_2,i_1)p(i_1,i)\right) = p(i,j)p^3(j,i)$$

so

$$p^3(i,j)p(j,i) = p(i,j)p^3(j,i).$$

For an arbitrary value of k, we would have

$$\sum_{i_k}\left(\cdots\sum_{i_1} p(i,i_1)p(i_1,i_2)\cdots p(i_k,j)p(j,i)\right) = p^{k+1}(i,j)p(j,i)$$

and

$$\sum_{i_k}\left(\cdots\sum_{i_1}p(i,j)p(j,i_k)\cdots p(i_2,i_1)p(i_1,i)\right)=p(i,j)p^{k+1}(j,i).$$

Now,

$$\lim_{k\to\infty}p^{k+1}(i,j)=\pi(j) \quad \text{and} \quad \lim_{k\to\infty}p^{k+1}(j,i)=\pi(i)$$

so

$$\pi(j)p(j,i)=\lim_{k\to\infty}p^{k+1}(i,j)p(j,i)=\lim_{k\to\infty}p^{k+1}(j,i)p(i,j)=\pi(i)p(i,j),$$

and the Markov chain is reversible.

Conversely, suppose that the Markov chain is reversible with stationary distribution π. We demonstrate the theorem in the case of $k = 3$.

We have

$$\pi(i_1)p(i_1,i_2)=\pi(i_2)p(i_2,i_1)$$

$$\pi(i_2)p(i_2,i_3)=\pi(i_3)p(i_3,i_2)$$

$$\pi(i_3)p(i_3,i_1)=\pi(i_1)p(i_1,i_3)$$

so

$$\begin{aligned}&\left[\pi(i_1)p(i_1,i_2)\right]\left[\pi(i_2)p(i_2,i_3)\right]\left[\pi(i_3)p(i_3,i_1)\right]\\&=\left[\pi(i_2)p(i_2,i_1)\right]\left[\pi(i_3)p(i_3,i_2)\right]\left[\pi(i_1)p(i_1,i_3)\right]\\&=\left[\pi(i_1)p(i_1,i_3)\right]\left[\pi(i_3)p(i_3,i_2)\right]\left[\pi(i_2)p(i_2,i_1)\right].\end{aligned}$$

Since $\pi(i_1)\pi(i_2)\pi(i_3) \neq 0$, we have

$$p(i_1,i_2)p(i_2,i_3)p(i_3,i_1)=p(i_1,i_3)p(i_3,i_2)p(i_2,i_1).$$

The result for an arbitrary k is done in a similar manner.

Example

The Markov chain whose transition matrix is

$$P = \begin{pmatrix} 1/2 & 1/3 & 1/6 \\ 1/6 & 1/2 & 1/3 \\ 1/3 & 1/6 & 1/2 \end{pmatrix}$$

is not reversible. The equilibrium state is

$$\pi = \left(\frac{1}{3}, \frac{1}{3}, \frac{1}{3}\right)$$

so

$$\pi(1)P_{12} = \frac{1}{3}\cdot\frac{1}{3} = \frac{1}{9} \quad \text{and} \quad \pi(2)P_{21} = \frac{1}{3}\cdot\frac{1}{6} = \frac{1}{18}$$

and the detailed balance equations are not satisfied.

Also

$$P_{12}P_{23}P_{31} = \frac{1}{3}\cdot\frac{1}{3}\cdot\frac{1}{3} \quad \text{and} \quad P_{13}P_{32}P_{21} = \frac{1}{6}\cdot\frac{1}{6}\cdot\frac{1}{6}$$

so Kolmogorov's conditions are not satisfied.

Example

The Ehrenfest models are a group of time-reversible Markov chains that were introduced in Chapter 2. In these models, there are M balls that are distributed into two urns. A ball is selected at random, and moved from the urn where it is, to the other. The state of the Markov chain is the number of balls in urn 1. Thus $S = \{0, 1, 2, \ldots, M\}$. The transition probabilities are

$$P_{00} = 0, \quad P_{01} = 1, \quad P_{MM} = 0, \quad P_{M,M-1} = 1,$$

$$P_{i,i+1} = \frac{M-i}{M}, \quad P_{i,i-1} = \frac{i}{M}, \quad i = 1,\ldots,M-1.$$

We use the detailed balance equations to find the equilibrium state. We have

$$\pi(0)P_{01} = \pi(0)\cdot 1 = \pi(0) \quad \text{and} \quad \pi(1)P_{10} = \pi(1)\cdot\frac{1}{M},$$

so

$$\pi(0) = \pi(1) \cdot \frac{1}{M} \quad \text{or} \quad \pi(1) = M\pi(0)$$

$$\pi(1)P_{12} = \pi(1) \cdot \frac{M-1}{M} \quad \text{and} \quad \pi(2)P_{21} = \pi(2) \cdot \frac{2}{M}, \text{ so}$$

$$\pi(1) \cdot \frac{M-1}{M} = \pi(2) \cdot \frac{2}{M} \quad \text{or} \quad \pi(2) = \frac{M-1}{2}\pi(1) = \frac{M-1}{2}M\pi(0)$$

$$\pi(2)P_{23} = \pi(2) \cdot \frac{M-2}{M} \quad \text{and} \quad \pi(3)P_{32} = \pi(3) \cdot \frac{3}{M}, \text{ so}$$

$$\pi(3) \cdot \frac{3}{M} = \pi(2) \cdot \frac{M-2}{M} \quad \text{or} \quad \pi(3) = \frac{M-2}{3}\pi(2) = \frac{M-2}{3} \cdot \frac{M-1}{2} \cdot M\pi(0).$$

A pattern is beginning to suggest itself. We have

$$\pi(1) = M\pi(0)$$

$$\pi(2) = \frac{M-1}{2}M\pi(0) = \frac{(M-1) \cdot M}{2}$$

$$\pi(3) = \frac{(M-2) \cdot (M-1) \cdot M}{3 \cdot 2}\pi(0).$$

This suggests that

$$\pi(4) = \frac{(M-3) \cdot (M-2) \cdot (M-1) \cdot M}{4 \cdot 3 \cdot 2}\pi(0),$$

and we leave it as an exercise to show this is indeed the case.

Now,

$$\binom{M}{k} = \frac{M!}{k!(M-k)!}, \quad \text{so} \quad \binom{M}{3} = \frac{M!}{3!(M-3)!} = \frac{M(M-1)(M-2)}{3!}$$

and

$$\binom{M}{4} = \frac{M(M-1)(M-2)(M-3)}{4!},$$

so we have for $k = 0, 1, 2, 3, 4$

$$\pi(k) = \binom{M}{k}\pi(0). \tag{2}$$

We leave it as an exercise to show equation (2) is true for $k = 0, 1, 2, \ldots, M$.

To determine $\pi(0)$, we use

$$\sum_{k=0}^{M}\pi(k) = \sum_{k=0}^{M}\binom{M}{k}\pi(0) = \pi(0)\sum_{k=0}^{M}\binom{M}{k} = \pi(0)2^M = 1$$

so

$$\pi(0) = \frac{1}{2^M} \quad \text{and} \quad \pi(k) = \frac{\binom{M}{k}}{2^M} \quad k = 0,1,2,\ldots,M.$$

Note that the Ehrenfest model is periodic of period 2.

We give the transition matrix and the equilibrium state for $M = 5$. We have

$$P = \begin{pmatrix} 0 & 1 & 0 & 0 & 0 & 0 \\ .2 & 0 & .8 & 0 & 0 & 0 \\ 0 & .4 & 0 & .6 & 0 & 0 \\ 0 & 0 & .6 & 0 & .4 & 0 \\ 0 & 0 & 0 & .8 & 0 & .2 \\ 0 & 0 & 0 & 0 & 1 & 0 \end{pmatrix}$$

$$\pi = \left(\frac{1}{32}, \frac{5}{32}, \frac{10}{32}, \frac{10}{32}, \frac{5}{32}, \frac{1}{32}\right).$$

We can check that

$$\left(\frac{1}{32}, \frac{5}{32}, \frac{10}{32}, \frac{10}{32}, \frac{5}{32}, \frac{1}{32}\right)\begin{pmatrix} 0 & 1 & 0 & 0 & 0 & 0 \\ .2 & 0 & .8 & 0 & 0 & 0 \\ 0 & .4 & 0 & .6 & 0 & 0 \\ 0 & 0 & .6 & 0 & .4 & 0 \\ 0 & 0 & 0 & .8 & 0 & .2 \\ 0 & 0 & 0 & 0 & 1 & 0 \end{pmatrix} = \left(\frac{1}{32}, \frac{5}{32}, \frac{10}{32}, \frac{10}{32}, \frac{5}{32}, \frac{1}{32}\right),$$

and we note that we cannot find the equilibrium state by raising P to a high power because it is periodic of period 2 (although the methods of Chapter 2 will work).

Example

In this example, we use the detailed balance equations to find the equilibrium state for a random walk with reflecting barriers. We consider the case where there are six states, 0, 1, 2, 3, 4, and 5, and the transition probabilities are

$$p(0,1)=1, \quad p(5,4)=1,$$

$$p(i,i+1)=p, \quad p(i,i-1)=q=1-p, 0<p,q<1, \quad i=1,2,3,4.$$

Let $\pi = (\pi_0, \pi_1, \pi_2, \pi_3, \pi_4, \pi_5)$ be the stationary state.

Using the detailed balance equations gives

$$\pi_0 p(0,1)=\pi_1 p(1,0) \quad \text{or} \quad \pi_0=\pi_1 q, \quad \text{so } \pi_1=\frac{1}{q}\pi_0$$

$$\pi_1 p(1,2)=\pi_2 p(2,1) \quad \text{or} \quad \pi_1 p=\pi_2 q, \quad \text{so } \pi_2=\frac{p}{q}\pi_1=\frac{p}{q}\frac{1}{q}\pi_0=\frac{p}{q^2}\pi_0$$

$$\pi_2 p(2,3)=\pi_3 p(3,2) \quad \text{or} \quad \pi_2 p=\pi_3 q, \quad \text{so } \pi_3=\frac{p}{q}\pi_2=\frac{p}{q}\frac{p}{q}\frac{1}{q}\pi_0=\frac{p^2}{q^3}\pi_0$$

$$\pi_3 p(3,4)=\pi_4 p(4,3) \quad \text{or} \quad \pi_3 p=\pi_4 q, \quad \text{so } \pi_4=\frac{p}{q}\pi_3=\frac{p}{q}\frac{p}{q}\frac{p}{q}\frac{1}{q}\pi_0=\frac{p^3}{q^4}\pi_0$$

$$\pi_4 p(4,5)=\pi_5 p(5,4) \quad \text{or} \quad \pi_4 p=\pi_5, \quad \text{so } \pi_5=p\pi_4=p\frac{p}{q}\frac{p}{q}\frac{p}{q}\frac{1}{q}\pi_0=\frac{p^4}{q^4}\pi_0.$$

We will want to extend this to N states, and this will be easier if we observe that

$$\pi_1=\frac{1}{q}\pi_0, \quad \pi_{n+1}=\frac{p}{q}\pi_n, \quad n=2,\dots,N-2, \quad \pi_{N-1}=p\pi_{N-2}.$$

We know

$$\sum_{n=0}^{N-1}\pi_n=1,$$

and if we know π_0, we can compute all the other π_i's.

In our example of six states, we have

$$\sum_{n=0}^{5} \pi_n = \pi_0 \left(1 + \frac{1}{q} + \frac{p}{q^2} + \frac{p^2}{q^3} + \frac{p^3}{q^4} + \frac{p^4}{q^4}\right) = 1.$$

To consider particular values, for $p = .6$ and $q = .4$, we get

$$\left(1 + \frac{1}{q} + \frac{p}{q^2} + \frac{p^2}{q^3} + \frac{p^3}{q^4} + \frac{p^4}{q^4}\right) = 26.375$$

so

$$\pi_0 = \frac{1}{26.375} = .0375,\ \pi_1 = .0948,\ \pi_2 = .1422,\ \pi_3 = .2133,\ \pi_4 = .3199,\ \pi_5 = .1919,$$

and one can check that

$$(.0375, .0948, .1422, .2133, .3199, .1919) \begin{pmatrix} 0 & 1 & 0 & 0 & 0 & 0 \\ .4 & 0 & .6 & 0 & 0 & 0 \\ 0 & .4 & 0 & .6 & 0 & 0 \\ 0 & 0 & .4 & 0 & .6 & 0 \\ 0 & 0 & 0 & .4 & 0 & .6 \\ 0 & 0 & 0 & 0 & 1 & 0 \end{pmatrix}$$

$$= (.0375, .0948, .1422, .2133, .3199, .1919).$$

Example

Next, we consider random walk on a semi-infinite line, {0, 1, 2, …}.

Consider the expression for sum of the six sites the stationary state:

$$\pi_0 \left(1 + \frac{1}{q} + \frac{p}{q^2} + \frac{p^2}{q^3} + \frac{p^3}{q^4} + \frac{p^4}{q^4}\right).$$

Ignoring the first and last summands in the parenthetical expression, we have

$$\frac{1}{q} + \frac{p}{q^2} + \frac{p^2}{q^3} + \frac{p^3}{q^4} = \frac{1}{q}\left(1 + \frac{p}{q} + \left(\frac{p}{q}\right)^2 + \left(\frac{p}{q}\right)^3\right),$$

and

$$1+\frac{p}{q}+\left(\frac{p}{q}\right)^2+\left(\frac{p}{q}\right)^3$$

is the beginning of a geometric series. If $p \geq q$, then the series diverges and the process is not positive recurrent. Suppose that

$$\frac{p}{q}=r<1.$$

Then,

$$\pi_0\left(1+\frac{1}{q}+\frac{p}{q^2}+\frac{p^2}{q^3}+\frac{p^3}{q^4}+\cdots\right)=\pi_0\left[1+\frac{1}{q}(1+r+r^2+r^3+\cdots)\right].$$

Now,

$$\frac{1}{q}\left(1+r+r^2+r^3+\cdots\right)=\frac{1}{q}\cdot\frac{1}{1-r}=\frac{1}{q}\cdot\frac{1}{1-\frac{p}{q}}=\frac{1}{q}\cdot\frac{q}{q-p}=\frac{1}{q-p}$$

and

$$1+\frac{1}{q}\left(1+r+r^2+r^3+\cdots\right)=1+\frac{1}{q-p}=\frac{q-p+1}{q-p}=\frac{q-p+p+q}{q-(1-q)}=\frac{2q}{2q-1}.$$

Note that $2q-1>0$, since $q>p$ and $p+q=1$.

So

$$\pi_0\left[1+\frac{1}{q}(1+r+r^2+r^3+\cdots)\right]=\pi_0\left(\frac{2q}{2q-1}\right)=1$$

and

$$\pi_0=\frac{2q-1}{2q}.$$

Also, for $i \geq 1$

$$\pi_i=\pi_0\frac{p^{i-1}}{q^i}.$$

For a specific example, suppose $q = .8$ and $p = .2$. The first nine entries in π (to five decimal places) are

$$\pi_0 = .375, \quad \pi_1 = .46875, \quad \pi_2 = .11719, \quad \pi_3 = .02930, \quad \pi_4 = .00732, \quad \pi_5 = .00183,$$
$$\pi_6 = .00046, \quad \pi_7 = .00011, \quad \pi_8 = .00003, \quad \pi_9 = .00001.$$

Raising the transition matrix to a high power is not feasible.

Random Walks on Weighted Graphs

An important class of reversible Markov chains is random walks on weighted graphs. In building the model, there is a one-to-one correspondence between the states of the Markov chain and the vertices of the graph, and we will refer to the states and the vertices with the same nomenclature. If P is the transition matrix for the Markov chain, then $P_{ij} > 0$ if and only if there is a directed edge from the vertex i to the vertex j. A directed edge from i to j will be assigned a weight, w_{ij}, that is related to P_{ij} in a way that we will describe shortly. We consider only the situation where $w_{ij} = w_{ji}$. The relationship between P_{ij} and w_{ij} is

$$P_{ij} = \frac{w_{ij}}{\sum_k w_{ik}}.$$

We now define a probability distribution on the Markov chain based on the graph weights that satisfies the detailed balance equations. Let s be the sum of the weights of all the edges; that is,

$$s = \sum_{i,j} w_{ij}.$$

Note that in the sum, both w_{ij} and w_{ji} are represented even though they have the same value.

Define the probability distribution on the Markov chain, $\pi = (\pi(1), \pi(2), \ldots)$, by

$$\pi(i) = \frac{\sum_k w_{ik}}{s}.$$

Example

Consider the graph in Figure 6.1.

Set $w_{ij} = 1$ if vertices i and j are connected. So

$$w_{12} = w_{21} = 1, \quad w_{23} = w_{32} = 1, \quad w_{24} = w_{42} = 1,$$

$$w_{34} = w_{43} = 1, \quad w_{ij} = 0 \text{ otherwise.}$$

Then

$$s = \sum_{i,j} w_{ij} = 8$$

so

$$\pi(1) = \frac{\sum_k w_{1k}}{s} = \frac{1}{8}, \quad \pi(2) = \frac{\sum_k w_{2k}}{s} = \frac{1}{8}, \quad \pi(3) = \frac{\sum_k w_{3k}}{s} = \frac{2}{8},$$

$$\pi(4) = \frac{\sum_k w_{4k}}{s} = \frac{2}{8}.$$

Then using

$$P_{ij} = \frac{w_{ij}}{\sum_k w_{ik}},$$

we have

$$P_{12} = 1, \quad P_{21} = P_{23} = P_{24} = \frac{1}{3}, \quad P_{32} = P_{34} = \frac{1}{2}, \quad P_{42} = P_{43} = \frac{1}{2},$$

$$P_{ij} = 0 \text{ otherwise.}$$

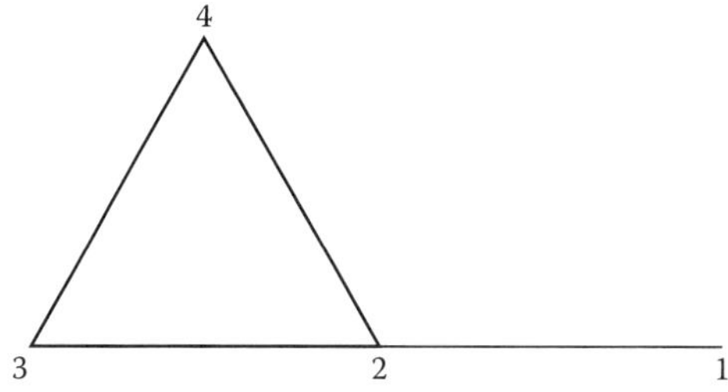

FIGURE 6.1
Weighted graph.

Thus,

$$P = \begin{pmatrix} 0 & 1 & 0 & 0 \\ 1/3 & 0 & 1/3 & 1/3 \\ 0 & 1/2 & 0 & 1/2 \\ 0 & 1/2 & 1/2 & 0 \end{pmatrix}.$$

Next, we show the detailed balance equations $\pi(i)P_{ij} = \pi(j)P_{ji}$ are satisfied. We have

$$\pi(1)P_{12} = \frac{1}{8}\cdot 1 = \frac{1}{8}, \quad \pi(2)P_{21} = \frac{3}{8}\cdot\frac{1}{3} = \frac{1}{8}$$

$$\pi(1)P_{13} = \frac{1}{8}\cdot 0 = 0, \quad \pi(3)P_{31} = \frac{2}{8}\cdot 0 = 0$$

$$\pi(1)P_{14} = \frac{1}{8}\cdot 0 = 0, \quad \pi(4)P_{41} = \frac{2}{8}\cdot 0 = 0$$

$$\pi(2)P_{23} = \frac{3}{8}\cdot\frac{1}{3} = \frac{1}{8}, \quad \pi(3)P_{32} = \frac{2}{8}\cdot\frac{1}{2} = \frac{1}{8}$$

$$\pi(2)P_{24} = \frac{3}{8}\cdot\frac{1}{3} = \frac{1}{8}, \quad \pi(4)P_{42} = \frac{2}{8}\cdot\frac{1}{2} = \frac{1}{8}$$

$$\pi(3)P_{34} = \frac{2}{8}\cdot\frac{1}{3} = \frac{1}{8}, \quad \pi(4)P_{43} = \frac{2}{8}\cdot\frac{1}{2} = \frac{1}{8}.$$

Next we note that

$$\pi P = \left(\frac{1}{8},\frac{3}{8},\frac{2}{8},\frac{2}{8}\right)\begin{pmatrix} 0 & 1 & 0 & 0 \\ 1/3 & 0 & 1/3 & 1/3 \\ 0 & 1/2 & 0 & 1/2 \\ 0 & 1/2 & 1/2 & 0 \end{pmatrix} = \left(\frac{1}{8},\frac{3}{8},\frac{2}{8},\frac{2}{8}\right) = \pi,$$

so π is a stationary state.

Discrete-Time Birth–Death Process as a Reversible Markov Chain

In a birth–death process, $P_{ij} = 0$ if $j \neq i \pm 1$. Suppose that π is the equilibrium state for a birth–death process. We will show that $\pi(i)P_{i,i+1} = \pi(i+1)P_{i+1,i}$. We construct a permeable membrane at $i + 1/2$ and view the equilibrium state as meaning the *probability flow* through the membrane must have a net change of zero. Stated mathematically, let K consist of the states $\{0, 1, \ldots, i\}$ and K^c consist of the states $\{i+1, i+2, \ldots\}$. If the process is in equilibrium, then

$$P(X_0 \in K) = P(X_1 \in K).$$

Now,

$$P(X_0 \in K) = P(X_0 \in K, X_1 \in K) + P(X_0 \in K, X_1 \in K^c)$$

and

$$P(X_1 \in K) = P(X_0 \in K, X_1 \in K) + P(X_0 \in K^c, X_1 \in K).$$

Since

$$P(X_0 \in K) = P(X_1 \in K),$$

this means

$$P(X_0 \in K, X_1 \in K^c) = P(X_0 \in K^c, X_1 \in K).$$

The event $(X_0 \in K, X_1 \in K^c)$ occurs exactly when the process is in state i at time 0 and state $i+1$ at time 1. In equilibrium, the probability of this event is $\pi(i)P_{i,i+1}$.

Likewise, the event $(X_0 \in K^c, X_1 \in K)$ occurs exactly when the process is in state $i+1$ at time 0 and state i at time 1. In equilibrium, the probability of this event is $\pi(i+1)P_{i+1,i}$. Thus

$$\pi(i)P_{i,i+1} = \pi(i+1)P_{i+1,i},$$

so we have a reversible Markov chain.

Continuous-Time Reversible Markov Chains

For continuous-time Markov chains, the infinitesimal generator G plays a role analogous to the transition matrix P for discrete-time Markov chains. The next theorem says how to determine whether a continuous-time Markov chain is reversible.

Theorem 6.5

Let $X(t)$ be an irreducible continuous-time Markov chain with infinitesimal generator G. Then, $X(t)$ is reversible if and only if there is a set of positive numbers $\{\pi_i\}$ for which

$$\pi_i G_{ij} = \pi_j G_{ji} \quad \text{for all } i \neq j, \quad \text{and} \quad \sum_i \pi_i = 1.$$

If this is the case, then $\pi = (\pi_1, \pi_2, \ldots)$ is the unique equilibrium distribution of the process.

These are the detailed balance equations for a continuous process.

Example

The M/M/1 queue is a continuous-time reversible Markov chain.

Recall the assumptions for the M/M/1 queue:

1. The states of the process are the number of customers in the system; that is, the customer being served plus those that are waiting.
2. Customers arrive according to a Poisson process with mean λ.
3. Service times are exponentially distributed with mean μ.

We have shown that the infinitesimal generator of the process is

$$G = \begin{pmatrix} -\lambda & \lambda & 0 & 0 & \cdots \\ \mu & -(\lambda+\mu) & \lambda & 0 & \cdots \\ 0 & \mu & -(\lambda+\mu) & \lambda & 0 \\ \vdots & \vdots & \vdots & \vdots & \vdots \end{pmatrix}$$

and that if $\lambda < \mu$ then there is a unique equilibrium state π, where

$$\pi(n) = \left(1 - \frac{\lambda}{\mu}\right)\left(\frac{\lambda}{\mu}\right)^n.$$

The detailed balance equations that we must check are

$$G_{i,i+1}\pi(i) = G_{i+1,i}\pi(i+1)$$

since $G_{i,j} = 0$ unless $|i-j| = 1$. Now,

$$G_{i,i+1}\pi(i) = \lambda\left(1 - \frac{\lambda}{\mu}\right)\left(\frac{\lambda}{\mu}\right)^i = \left(1 - \frac{\lambda}{\mu}\right)\frac{\lambda^{i+1}}{\mu^i}$$

and

$$G_{i+1,i}\pi(i+1)=\mu\left(1-\frac{\lambda}{\mu}\right)\left(\frac{\lambda}{\mu}\right)^{i+1}=\left(1-\frac{\lambda}{\mu}\right)\frac{\lambda^{i+1}}{\mu^{i}},$$

so the detailed balance equations are satisfied and so the Markov chain is reversible.

Example

Suppose we have two generators that act independently. Each fails and is repaired at an exponential rate. The first fails at the rate α_1 and is repaired at the rate β_1. The second fails at the rate α_2 and is repaired at the rate β_2. We will show that this does not give a reversible Markov chain—a result that should not be surprising.

We denote the states as follows:

1—Both generators are functioning.

2—The first generator is functioning; the second is down.

3—The first generator is down; the second is functioning.

4—Both generators are down.

The infinitesimal generator for the process is

$$G=\begin{pmatrix}-(\alpha_1+\alpha_2) & \alpha_1 & \alpha_2 & 0\\ \beta_1 & -(\beta_1+\alpha_2) & 0 & \alpha_2\\ \beta_2 & 0 & -(\alpha_1+\beta_2) & \alpha_1\\ 0 & \beta_1 & \beta_2 & -(\beta_1+\beta_2)\end{pmatrix}.$$

The equilibrium state is the probability vector $\hat{x}=(x_1,x_2,x_3,x_4)$ for which

$$\hat{x}G=\hat{0}.$$

Now,

$$(x_1,x_2,x_3,x_4)\begin{pmatrix}-(\alpha_1+\alpha_2) & \alpha_1 & \alpha_2 & 0\\ \beta_1 & -(\beta_1+\alpha_2) & 0 & \alpha_2\\ \beta_2 & 0 & -(\alpha_1+\beta_2) & \alpha_1\\ 0 & \beta_1 & \beta_2 & -(\beta_1+\beta_2)\end{pmatrix}$$

$$=(-(\alpha_1+\alpha_2)x_1+\beta_1x_2+\beta_2x_3,\ \ \alpha_1x_1-(\beta_1+\alpha_2)x_2+\beta_1x_4,$$

$$\alpha_2x_1-(\alpha_1+\beta_2)x_3+\beta_2x_4,\ \ \alpha_2x_2+\alpha_1x_3-(\beta_1+\beta_2)x_4).$$

Thus, we must solve

$$-(\alpha_1+\alpha_2)x_1+\beta_1x_2+\beta_2x_3=0$$

$$\alpha_1 x_1 - (\beta_1 + \alpha_2) x_2 + \beta_1 x_4 = 0$$

$$\alpha_2 x_1 - (\alpha_1 + \beta_2) x_3 + \beta_2 x_4 = 0$$

$$\alpha_2 x_2 + \alpha_1 x_3 - (\beta_1 + \beta_2) x_4 = 0$$

$$x_1 + x_2 + x_3 + x_4 = 1.$$

The last equation is necessary because $\hat{x}$ is a probability vector.

If we set $\alpha_1 = 4$, $\alpha_2 = 6$, $\beta_1 = 1$, and $\beta_2 = 2$, then

$$\hat{x} = \left(\frac{17}{274}, \frac{32}{274}, \frac{69}{274}, \frac{156}{274}, \right).$$

To determine whether the chain is reversible, we check the detailed balance equations. Now,

$$G_{12} x_1 = \alpha_1 x_1 = 4 \cdot \frac{17}{274} = \frac{68}{274}$$

$$G_{21} x_2 = \beta_1 x_2 = 1 \cdot \frac{32}{274} = \frac{32}{274},$$

so the chain is not reversible.

Exercises

6.1 Show that the irreducible Markov chain whose transition matrix is

$$P = \begin{pmatrix} 1/8 & 3/8 & 0 & 1/2 \\ 1/2 & 1/8 & 3/8 & 0 \\ 0 & 1/2 & 1/8 & 3/8 \\ 3/8 & 0 & 1/2 & 1/8 \end{pmatrix}$$

is not reversible.

6.2 (a) We have 2 containers, 3 white balls, and 3 green balls. The 6 balls are divided so that each container has 3 balls. One ball is chosen from each container and moved to the other container. Construct a Markov chain where the states are the number of white balls in the first container.

(i) Construct the transition matrix for the Markov process.

(ii) Find the equilibrium state for the Markov process.

(iii) Show that the Markov process is reversible.

(b) Repeat part (a) when there are 4 white balls and 4 black balls.

(c) Based on your results for parts (a) and (b), make a conjecture about the equilibrium state when there are M white balls and M black balls.

Construct the transition matrix, verify your conjecture, and verify that the Markov chain is reversible.

6.3 Suppose that $\{X\}$ is a finite state Markov chain with transition matrix P. Show that if P is symmetric (i.e., $P(i,j) = P(j,i)$), then $\{X\}$ is reversible and the uniform probability distribution is the equilibrium state.

6.4 For the Markov chain whose transition matrix is

$$P = \begin{pmatrix} \frac{1}{2} & 0 & \frac{1}{2} \\ \frac{1}{3} & \frac{1}{6} & \frac{1}{2} \\ \frac{1}{4} & \frac{1}{4} & \frac{1}{2} \end{pmatrix},$$

(a) Find the equilibrium state.

(b) Find the transition matrix for the reversed Markov chain.

6.5 (a) Find the equilibrium state for the Markov chain whose transition matrix is

$$P = \begin{pmatrix} 1-\alpha & \alpha \\ \beta & 1-\beta \end{pmatrix} \quad 0 < \alpha, \beta < 1.$$

(b) Show that the Markov chain is reversible.

6.6 For the weighted graphs given in Figure 6.2,

(a) Find the transition matrix of the associated Markov chain.

(b) Find the equilibrium state of the Markov chain.

(c) Verify that the Markov chain is reversible.

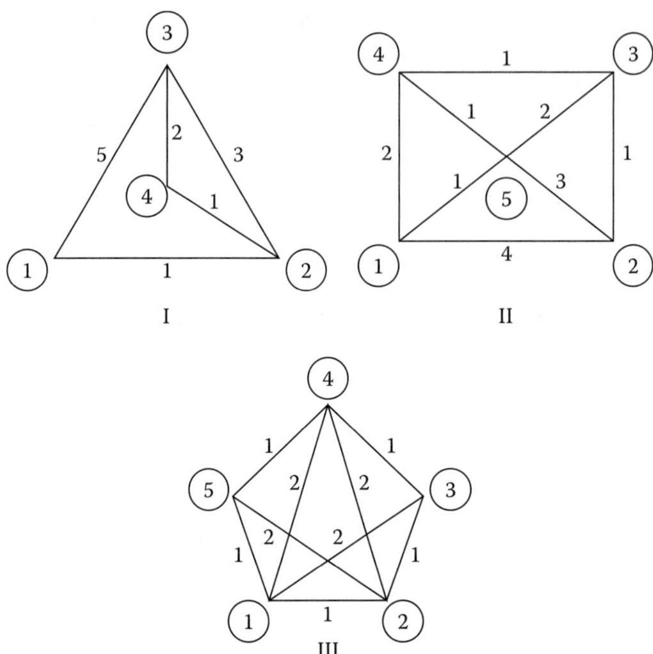

FIGURE 6.2
Weighted graph for Problem 6.6.

Bibliography

Anderson, D.F. 2014. Stochastic methods for biology, University of Wisconsin, Madison, WI. http://www.math.wisc.edu/~anderson/605F11/Notes/StochBio (accessed on October 17, 2014).

Breiman, L., *Probability*, Addison-Wesley, Reading, MA, 1968.

Feller, W., An Introduction to Probability Theory and Its Applications, vol. I, 3rd edn., Wiley, New York, 1968.

Gallager, R. 2014. Discrete stochastic processes, MIT; http://en.wikipedia.org/wiki/Cambridge,_Massachusetts Cambridge, MA. http://www.mit.edu/courses/electrical-engineering-and-computer-science/6-262-discrete-stochastic-processes-spring-2011/course-notes/MIT6_262S11_chap02.pdf (accessed on October 17, 2014).

Hailovic, A. 2014. KTH Royal Institute of Technology, Stockholm, Sweden. http://ingforum.haninge.kth.se/armin/ALLA_KURSER/KOTEORI/EXER/repet2b.pdf (Possible source for exercises) (accessed on October 17, 2014).

http://www.math.wisc.edu/~anderson/605F13/Notes/StochBio.pdf.

Kemeny, J. and J. Snell, *Finite Markov Chains*, Van Nostrand, Princeton, NJ, 1960.

Lawler, G., *Introduction to Stochastic Processes*, Chapman & Hall CRC, Boca Raton, FL, 2006.

Li, J. 2014. Stochastic modeling, Pennsylvania State University, State College, PA. http://sites.stat.psu.edu/~jiali/course/stat416/notes/ (accessed on October 17, 2014).

Probability for comp science (Queuing Theory). http://www.public.iastate.edu/~riczw/stat330s11/lecture/lec22.pdf (accessed on October 17, 2014).

Romano, J. 2014. Introduction to stochastic processes, Stanford University, Stanford, CA. http://www.stanford.edu/class/stat217/ (accessed on October 17, 2014).

Ross, S., *Introduction to Probability Models*, Academic Press, New York, 1989.

Siegrist, K. 2014. Universal laboratories in probability and statistics, University of Alabama, Huntsville, AL. http://www.math.uah.edu/stat/prob/ (accessed on October 17, 2014).

Sigmon, K. 2014. Introduction to renewal theory, Columbia University, New York. http://www.columbia.edu/~ks20/stochastic-I/stochastic-I-RRT.pdf (accessed on October 17, 2014).

Takahara, G. 2014. Stochastic processes, Queens University, Kingston, Ontario, Canada. http://www.mast.queensu.ca/~stat455/lecturenotes/lecturenotes.shtml (accessed on October 17, 2014).

Index

M

N

O

P